Engineering and Architecting Multidisciplinary Systems

Volume 4

EVALUATION AND PROOF

OF THE SYSTEM

ALAIN FAISANDIER

Sinergy'Com

Practical Guidelines

COPYRIGHT

Copyright © 2022 Sinergy'Com

All rights reserved. No part of this publication may be reproduced, stored in a retrieval system, or transmitted in any form or by any means, electronic, mechanical, photocopying, scanning, or otherwise, without prior written permission of the publisher.

Permissions may be sought directly from publisher Sinergy'Com.

Sinergy'Com
2 chemin de la Serre
31450 BELBERAUD - France

contact@sinergycom.net

ISBN: 979-10-91699-10-5

This book is also available in e-book pdf format (ISBN: 979-10-91699-11-2).

FOREWORD

Purpose of this volume

The economic, security and societal stakes of systems created by human beings are such that their final validation must be acquired before the reception. These challenges alone justify the implementation of evaluation, verification and validation activities at the earliest and throughout the development. Verification and validation costs can become so onerous that it is necessary to rationalize the approach to produce the expected effects with maximum efficiency.

Justification, validation and verification activities aim to give confidence to all involved parties, recognizing that the system (product, service, organization) complies with the requirements and respects the expected architectural and design properties.

The set of evaluation, justification, verification and validation activities represents the engineering support node, which all other development activities must use in order to make relevant technical decisions.

Faced with any problem, the natural approach of every individual is to find a solution as quickly as possible. Because the mere fact of posing a problem, or a question, creates a vacuum; but nature tries to fill this gap as quickly as possible in order to regain a certain stable state of equilibrium, in which there is no longer any problem.

An engineer is theoretically capable of finding a solution to a technical issue or problem that has never been resolved before. Without knowledge of engineering, and without training in all its components, the reflex of the engineer is the same as any other individual. Each individual seeks knowledge stored in their own head, or the objects that surround it, elements to define a solution, and to materialize it quickly. This does not necessarily involve extensive evaluation and verification of the feasibility or appropriateness of this solution.

Any experienced engineer should know the process of creation in detail, i.e. the engineering discipline, in order to answer the problem or the question asked, in a progressive and sure way. To understand the engineering approach, here is a metaphor: the problem or question is like a black tunnel without light, or a labyrinth, into which human beings penetrate, but which potentially contains obstacles, traps and/or several galleries. Some galleries lead to more or less easy or adapted exits, others have no exit; the "right" exit is the "right" solution to the problem or the "right" answer to the question asked. The engineering approach consists of lighting a few lights in the tunnel, that is to acquaint oneself with the present environment, to draw up plans to overcome obstacles, to evaluate the solidity of the walls of the galleries, to consolidate them, to check the progress, etc., and all this step by step, until the "right" exit is identified. The engineering approach is not natural, instinctive or intuitive; it is not just common sense. It is made up of concepts, principles and techniques that must be known, learned and used.

The "Engineering and Architecture of Multidisciplinary Systems" series develops the concepts, principles and techniques of engineering:

- Volume 1, System Notion and Engineering of Systems [Faisandier1 2015], presents an overview of engineering, a sort of summary, a map or a plan to set the benchmarks, to show the approaches and the concepts in respect of each other.

- Volume 2, Systems Opportunities and Requirements [Faisandier2 2013], explains how a need, opportunity, problem or issue is identified by characterizing the environment, developing stakeholder requirements repositories, expected requirements or the characteristics applicable to a potential system that is able to respond to the issue or opportunity.

- Volume 3, Systems Architecture and Design [Faisandier3 2013], explains how alternative architectural and design solutions are defined to satisfy the need or the opportunity, starting from the requirements repository, on the one hand, and from the patterns or generic models corresponding to the problem concerned, on the other hand.

- This Volume 4, Evaluation and Proof of the System, explains how alternative solutions are evaluated and verified, and how the definition of the selected solution, as well as its tangible realization, is verified, validated and justified.

Evaluation, justification, verification and validation have in common the notion of comparison. Comparison of alternatives in respect of the criteria to ensure selection of the most relevant solution (evaluation and justification); comparison of the selected solution in respect of technical characteristics references (verification) and operational references (validation).

Evaluation, justification, verification and validation activities are an integral part of engineering. But they are transversal of the activities of the definition of the requirements and the definition of the architectural and design solutions. More generally, they are transversal to development activities (definition of concepts, definition of the system, realization and integration of the tangible constituents of the system); and more generally they are crosscutting across all activities in the life cycle of the system, in order to obtain and maintain over time a product, service or organization that complies with the needs. These activities must be carried out in coordination with each of the other engineering activities, gradually, and not only at the end of the implementation of the system.

In order to show the articulation of the verification and validation activities on the whole development of the system, this volume also deals with the integration of the system; more precisely the integration of the implemented system elements (constituents) of the product, service or organization corresponding to the system-of-interest.

Abstract content

The present volume is dedicated to the evaluation and proof of the system. The following topics are covered:

- The **evaluation** and **justification** of alternative solutions of the system studied, whether represented by a set of expected characteristics (requirements references), or architectural and/or design properties and models. The evaluations are carried out against objective and subjective criteria, and/or expert opinions, which depend on the type of system. This evaluation activity is called "System Analysis" and/or sometimes "System Assessment" in systems engineering standards.

- **Verification** of the engineering entities relating to the system (requirements, functions, input-output flows, tangible system elements, interfaces, integration aggregates, system, etc.); the verifications are done against their own definition reference.

- **Validation** of the same engineering entities relating to the system (requirements, functions, input-output flows, tangible system elements, interfaces, integration aggregates, system, etc.); the validations are done against their own utilization references.

- **Integration** of the tangible components of the system (implemented system elements) to build the expected product, service or organization, following an effective approach.

Efficiency is achieved through optimal and systemic coordination of assembly, verification and validation activities. This coordination is called **integration, verification, and validation strategy**.

Content of other volumes

The series of books entitled "Engineering and Architecting Multidisciplinary Systems" is composed of several volumes written to care about coherence of terminology and between mentioned notions, concepts and principles. The first four volumes are in line with the following collective works: the guide of the body of knowledge SEBoK [SEBoK 2016], the INCOSE Handbook version 4 [INCOSE 2015], the standard ISO/IEC/IEEE 15288:2015 [ISO15288], the standards ISO/IEC 24748-1:2016 Guide for life cycle management [ISO 24748-1], ISO/IEC 24748-2:2016 Guide to the application of ISO/IEC 15288:2015 [ISO 24748-2], and ISO/IEC 24748-6:2016 System integration engineering [ISO 24748-6]. Titles and summarized contents of the volumes are as follow:

- Volume 1 - System Notion and Engineering of Systems
    - Notion of System; System vision; Principles for engineering a system; Means for engineering systems; Relationships to Project Management and technologies engineering; System life cycle management

- Volume 2 - Systems Opportunities and Requirements
    - Mission Analysis; Business Opportunity Analysis; Operational and Technological Concepts; Stakeholder Requirements Definition; System Requirements Definition

- Volume 3 - Systems Architecture and Design
    - System Logical Architecture Definition; System Physical Architecture Definition; Modelling and patterns; System Design and System Element Design

- Volume 4 - Evaluation and Proof of the System (the present volume)
    - System Analysis (system properties assessment); System Verification; System Validation; System Integration

- Volume 5 - Engineering of Safe, Secure and Resilient Systems
    - Immunity, integrity, harmless aspects applied to systems; Extension of Systems Engineering to Dependability Engineering and integrated approach; FDIR approach and redundancy patterns; Assessment of dependable properties

- Volume 6 - System of Interest Life Management
    - Implementation Technologies; Integration, deployment, maintenance, disposal as enabling systems; System life cycle synthesis

- Volume 7 - Systems Engineering Ontology
    - Principles to establish a meta-data model; Semantic bus and engineering ontology; Detailed description of the generic engineering meta-data model used in other volumes

The practical guidelines of this series of books are intended primarily for professionals, either customer or supplier, who want to understand and apply daily engineering of systems; in other words for those who will have to define operational concepts, express stakeholder and system requirements, define candidate architectures, design and validate complex or multidisciplinary systems. In these guidelines they will find methods, processes, modelling and analysis techniques, reasoning elements, useful recommendations for application, as well as case studies or examples to use this approach.

Foreword

Other people having to understand or to exchange with system engineers will find simple and concrete explanations. They will then be able to judiciously apply certain precepts and concepts to their own job. For example, they can be responsible for architecture, design, integration, verification and validation, program and project managers, technological study engineers, service developers; or they can be in charge of organising sets or complex enterprises whatever the sector (technical, commercial, military, academia, government, tertiary).

These guidelines are most useful to University professors who desire teaching Engineering of Systems, Architecting of Multidisciplinary Systems, to university students, and to researchers, as a certain number of topics discussed in these books could possibly lead to more extensive studies, as well as studies which have not yet been tackled by the engineering community.

These guidelines describe the fundamentals and the means that can be then peacefully deepened with discernment through the literature about these subjects. They include definitions, descriptions, discussions; examples and case studies illustrate practices. The case studies explain step by step how to perform the activities and tasks. Readers can directly transpose the step-by-step and use provided templates for their current or future projects.

Sources and motivation

As the author has a long career in engineering disciplines, is a member of professional associations like the International Council On Systems Engineering (INCOSE), and in particular contributes to Working Groups for international standardisation (ISO-IEC) on the subject, a certain number of concepts, activities, methods, modelling techniques and advice have been selected to be presented here. They may be found in other books, publications, documents, and training existing prior to the publication of this series of books. The Guide to Systems Engineering Body of Knowledge version 1.0, published at the end of 2012, now continuously updated, and accessible on the web, for which the author is contributing, also contains such materials; but more explanations and details are discussed and developed in the volumes of this series of books.

The content of this series of books comes from experiences of practitioners who were facing situations during their professional jobs (and even throughout their lives) for which no, or very few, known methods existed. The author of this book has not invented all the described concepts and principles; the majority of them existed from a long time ago, but here they are clarified and presented as a consistent set to be applied immediately and simply. Interesting references and citations from standards and books are provided, but in limited number. The reader will find here basic practices about *What-to-do* using a standard vocabulary; but the interest of these books resides more in the details of *How-to-do*.

The main sources come from guidelines written previously for several companies and organizations, and from training materials that the author has used over the last twenty years to teach Systems Engineering in industry, government and academia concerning more than five thousand people.

The original feature of this series of books is to gather together a whole coherent set of practices, guidelines, recommendations and templates directly applicable to projects.

The author has not stopped to harmonise interesting studies and works about concepts and principles such as those used here; to present them clearly in simple terms, while these elements are also presented separately in other books, documents or standards.

"Coherence" between concerned notions is the key term of this series of books.

Practical means used

Colour code

Words coloured in brown are elements of the generic engineering meta-data model. Verbs coloured in blue express relationships between these elements.

Terms coloured **black** or **blue** and with bold characters represent focal points inside a section or a sub section of text.

Texts are written in UK English.

Acknowledgements

The author of these guides has spent a lot of time in consultancy and teaching engineering, architecting and designing disciplines in international context during his career. Time and time again he has been encouraged by many people to publish what he taught and discussed to benefit practitioners in companies, academia and government organisations.

I am pleased to acknowledge my colleagues who had philosophical, technical, and practical discussions with me, and those who influenced me in particular for this volume, including Therese Renard. A special thank you to Zoe Miller who corrected the English language.

I would also like to thank sponsors of several companies and government organisations who contracted with MAP systeme, the company I managed, for producing methodological guides which prefigure some parts of the present book; in particular PSA Peugeot Citroën and VALEO in the automotive sector, BioMerieux Laboratories in the biomedical sector, SNECMA (SAFRAN Group) in the aeronautic sector, and NAGRA Vision in the television and security information domain.

A special thank you to the Direction Générale de l'Armement (DGA) of the French Ministry of Defence for having sponsored the projects KIMONO-OMOTESC (five million Euros) which enabled specifying, designing, realising and delivery of the tool MDWorkbench ®. I have been the Project and Technical Manager for the development of this Systems Engineering workbench using the ISO/IEC/IEEE 15288 processes as basis. Engineering and defining as a system this workbench which supports Systems Engineering processes including modelling techniques, taught and questioned me a lot about the practice of the discipline; in particular with the ontology and the necessary central generic engineering meta-data model.

About the author

Alain Faisandier is director and founder of MAP systeme (1996), trainer and international senior expert in Systems Engineering. During his forty-year career, he experienced engineering of complex software and systems for large, medium and small projects, in national and international contexts, through requirements definition, architecture, design, integration, verification, validation, and dependability. He has also practiced project management, quality assurance, as well as developer of requirements, architecture, design, and verification-validation methods for systems.

During the 1970's, he worked for the International Telegraph & Telephone Company (ITT) in the context of public telecommunication systems as designer, integrator, and technical manager. During the 1980's and 1990's, he worked for the European space industry, participating in the methodology development of major European space programs within MATRA MARCONI Space Company that became Astrium, an EADS Company. At this time he studied and worked with ESA

(European Space Agency) and NASA (USA National Aeronautics and Space Administration) standards in particular.

Since 1998 he participates actively in the deployment of Systems Engineering for large companies and organisations in defence, aeronautical, automotive, telecommunication, railway transportation, biomedical, medical, power / energy, and infrastructures sectors. This deployment includes a combination of long term actions, such as training, project coaching, writing guidelines, and tooling. These actions are carried out in multi-disciplinary, multi-cultural and international contexts.

He is also an academic professor in Systems Engineering in several French High schools and Universities for students and also for professors and for the benefit of researchers.

He has been co-chairman of the INCOSE Standard Technical Committee, 2000 - 2005, and head of the French Delegation for ISO/IEC 15288 standard development and its application guide. From 2009 till 2011 he was Technical Director of the Association Française d'Ingénierie Système - AFIS (an affiliate association of INCOSE). Finally he participates to the international project "Body of Knowledge and Curriculum to Advance Systems Engineering" (BKCASE) as co-author of the "Guide to the Systems Engineering Body of Knowledge" [SEBoK 2016], as well as the update of the INCOSE Systems Engineering Handbook [INCOSE 2015], and of the standard ISO/IEC 15288 - System life cycle processes [ISO 15288].

Alain Faisandier - MAP systeme Managing Director

Templates of technical documents

Word© format of templates provided in annexes are available for download at www.mapsysteme.com heading "Books".

Errare humanum est, sed perseverare diabolicum - Seneque and/or Saint Augustin

(Error is human, but to persist is diabolical.)

CONTENT

FOREWORD 1

1 INTRODUCTION 11
2 TERMINOLOGY 14
   2.1   Terms and definitions 14
   2.2   Abbreviations 21
3 REMINDERS AND EXTENSIONS ABOUT FUNDAMENTALS FOR THE DEVELOPMENT OF THE SYSTEM 23
   3.1   Notion of System 24
   3.2   Processes to define the concepts and the system 30
   3.3   Processes for developing the system 34
      3.3.1   Notion of system-block 34
      3.3.2   Development approach using processes and system-blocks 35
      3.3.3   System properties assessment approach 38
      3.3.4   Verification and validation approach 39
   3.4   Ontology for engineering a system 40
      3.4.1   Ontology for development of the system 40
      3.4.2   Ontology for the definition of the system 42
   3.5   Intellectual creation and origin of errors 44
      3.5.1   A mental behavioural model 44
      3.5.2   Creation mechanism and systems engineering activities 45
      3.5.3   Relevance of evaluation, verification and validation approaches with respect to human "reliability" 48

4 SYSTEM PROPERTIES ASSESSMENT 53
   4.1   Assessment concepts and principles 54
      4.1.1   Trade-off studies 54
      4.1.2   Cost analysis 58
      4.1.3   Risk analysis 64
      4.1.4   Effectiveness analyses 69
      4.1.5   Assessment criteria 72
      4.1.6   Classification of candidate solutions and optimization 74
   4.2   Process approach 77
      4.2.1   Location of the process in the development cycle 77
      4.2.2   Purpose of the process, inputs and outputs 78
      4.2.3   Activities of the process 80
      4.2.4   Ontology elements 83
      4.2.5   Verification and validation of evaluations 85
      4.2.6   Artefacts - documentation 86
   4.3   Practice 87
      4.3.1   Applicable models 87
      4.3.2   Experts opinion and interviews 92
      4.3.3   Assessment example 95
      4.3.4   Recommendations and frequently asked questions 96

5 VERIFICATION AND VALIDATION OF THE SYSTEM 99
   5.1   Concepts, principles, approaches 100
      5.1.1   Definitions 100
      5.1.2   Notion of Verification Action and Validation Action 103
      5.1.3   Verification and validation methods and techniques 107

	5.1.4	Overview of verifications and validations	114
	5.1.5	Verification and validation strategy	121
5.2	Process approach		130
	5.2.1	Verification process	130
	5.2.2	Validation process	140
5.3	Practice		150
	5.3.1	Traceability and justification matrix	150
	5.3.2	Inspection technique	156
	5.3.3	End-of-activity review technique	158
	5.3.4	Test technique	162
	5.3.5	Organization of verification and validation activities	186
	5.3.6	Recommendations and frequently asked questions	187

6 INTEGRATION OF SYSTEM COMPONENTS ... 191

6.1	Concepts and approaches for integration		192
	6.1.1	Definitions and concepts	192
	6.1.2	Integration approaches	196
	6.1.3	Integration strategy	200
	6.1.4	Relationship of integration with verification and validation	208
6.2	Process approach		212
	6.2.1	Location of the process	212
	6.2.2	Purpose, inputs and outputs of the process	213
	6.2.3	Activities of the process	214
	6.2.4	Ontology elements	218
	6.2.5	Artefacts - documentation	220
6.3	Practice		221
	6.3.1	Components heterogeneity and interfacing complexity	221
	6.3.2	Notion of Integration-Enabling-System	224
	6.3.3	Interfacing with other development processes	226
	6.3.4	Searching for and clearing defects	229
	6.3.5	Recommendations and frequently asked questions	234

7 SOME REFERENCES ... 237

8 ANNEXES ... 239

8.1	System Justification Document - Template and guidelines	239
8.2	System Verification and Validation Plan - Template and guidelines	247
8.3	Justification Matrix	251
8.4	Verification - Validation Action Sheet	265
8.5	Verification - Validation Procedure Sheet	266
8.6	Anomaly or Non-conformance Sheet	267
8.7	Inspection Sheet	268
8.8	Problem Sheet	268
8.9	System Integration Plan - Template and guidelines	269
8.10	Integration Aggregate Definition Sheet	273
8.11	Integration Procedure	274

FIGURES

Figure 1 - A model of the definition of system ... 24
Figure 2 - Hierarchical breakdown of the system-of-interest (SBS) ... 27
Figure 3 - Encapsulation of systems with their connections ... 28
Figure 4 - Encapsulation of the system-of-interest in its context ... 28
Figure 5 - Recursion of the definition of systems ... 29

Figure 6 - Iteration and recursion of definition processes 31
Figure 7 - Transition towards System Element implementation processes 33
Figure 8 - Notion of system-block for development purpose 34
Figure 9 - System development processes 35
Figure 10 - System Element implementation processes 37
Figure 11 - Assessment process and approach 38
Figure 12 - Verification and validation processes and approach 39
Figure 13 - Simplified view of the engineering meta-model for development 41
Figure 14 - Simplified view of the engineering meta-model for definition 42
Figure 15 - A knowledge acquisition model 44
Figure 16 - Intellectual creation / conceptualization mechanism 46
Figure 17 - Origin of observable failures 50
Figure 18 - Evolution of quantifications 56
Figure 19 - Method for conducting trade-off studies 57
Figure 20 - Example of life cycle cost partition in time 59
Figure 21 - Example of a conversion cost / estimate scale 60
Figure 22 - Paradigm of the risk 64
Figure 23 - Example of a severity logarithmic scale for a planning risk 66
Figure 24 - Example of a risk acceptability diagram 67
Figure 25 - Table with nine cases of risks criticality 67
Figure 26 - Delimitation of the solution domain 72
Figure 27 - Weighed criteria tree 73
Figure 28 - Example of a multi-criteria model 73
Figure 29 - Location of the assessment process in the development 77
Figure 30 - Support aspect of the assessment process 79
Figure 31 - Activities of the assessment process 80
Figure 32 - Relationships between engineering meta-data related to assessment 84
Figure 33 - Example of a cost learning curve 90
Figure 34 - Example of a beta curve with 50% and 81% profiles 90
Figure 35 - Necessary elements for a verification action 103
Figure 36 - Necessary elements for a validation action 104
Figure 37 - Position of techniques: static / dynamic and theoretical / experimental 107
Figure 38 - When to verify and validate? 115
Figure 39 - Internal verifications for activities 116
Figure 40 - End-of-activity validations 117
Figure 41 - Dynamic experimental verifications and validations 119
Figure 42 - Verification and validation of the system layer by layer 120
Figure 43 - Location of the verification process in the development 130
Figure 44 - Activities of the verification process 133
Figure 45 - Relationships between the engineering meta-data related to verification 138
Figure 46 - Location of the validation process in the development 140
Figure 47 - Activities of the validation process 143
Figure 48 - Relationships between the engineering meta-data related to validation 148
Figure 49 - Principle for establishing the Justification Matrix 152
Figure 50 - Execution of the inspection 157
Figure 51 - Execution of the end-of-activity review 160
Figure 52 - Working of an entity 162
Figure 53 - Behaviour of an entity 162
Figure 54 - Necessary elements for a generic test 163
Figure 55 - Necessary elements to define a generic test case 167
Figure 56 - Elaboration of test data sets 168
Figure 57 - Vending machine; States-Transitions diagram 170
Figure 58 - Example of a States-Transitions diagram 173
Figure 59 - Example of a States-Transitions matrix 173
Figure 60 - Example of transactional flows - Cash Machine 175
Figure 61 - Activities of tests implementation 177

Figure 62 - Development of a test procedure 179
Figure 63 - Sequence of tests 180
Figure 64 - Preparation of the target product for testing 181
Figure 65 - Run of tests sessions 182
Figure 66 - Analysis of the anomaly 184
Figure 67 - Decision to stop or continue testing 185
Figure 68 - Scope of system integration 193
Figure 69 - Integration by layers of system 198
Figure 70 - Incremental integration 202
Figure 71 - Subsets integration (aggregates) 203
Figure 72 - Top-down integration 204
Figure 73 - Bottom-up integration 205
Figure 74 - Assembly techniques 206
Figure 75 - Example of coupling matrices to define integration aggregates 207
Figure 76 - Coupling matrix to define test cases for input / output flows 208
Figure 77 - Location of the integration process in the development 212
Figure 78 - Activities of the integration process 214
Figure 79 - Relationships between engineering meta-data related to integration 219
Figure 80 - Integration by similar technologies (read from bottom to top) 222
Figure 81 - Relationship between engineering of the system and engineering of technologies 223
Figure 82 - Interactions between the processes used for the SOI and the IES 225
Figure 83 - Search for and clear defect 229
Figure 84 - Test cases to re-perform to prove non-regression 233

TABLES

Table 1 - Life cycle cost items 61
Table 2 - Estimate of the "cost" of the functionalities of the system 62
Table 3 - Table of marks of candidate solutions 74
Table 4 - Comparative table of the score of candidate solutions 75
Table 5 - Main engineering meta-data related to assessment 84
Table 6 - Pitfalls about assessment 96
Table 7 - Good practices about assessment 96
Table 8 - Main engineering meta-data related to verification 137
Table 9 - Main engineering meta-data related to validation 147
Table 10 - Tests mapping 186
Table 11 - Main engineering meta-data related to integration 219

F A Q

System properties assessment 97
Verification and validation of the system 188
System integration 235

1 INTRODUCTION

> **Evaluate, justify, verify, validate**

For the neophyte in engineering, the terms evaluate, justify, verify and validate are more or less synonymous, even enigmatic. These action verbs cover more or less complex activities, called support activities, common to other engineering activities. As described in Volume 1 of this series of books, System Notion and Engineering of Systems [Faisandier1 2015] (Chapter 5.4.2, Categories of processes and maps), engineering consists of creating solutions to problems or opportunities identified in one given context. In the first approach, creating means analysing the contextual situation(s), understanding the need or the opportunity, imagining possible solutions, and presenting them in a language understandable to the stakeholders.

But any imagined solution is not necessarily feasible, because reality has its own laws, and technological means are not necessarily available at the right time. The numerous drawings and models of Leonardo da Vinci (1452-1519) attest to this fact: it took a few centuries to implement and build submarines, aircrafts, helicopters or tanks. Before realizing solutions, it is necessary to characterize the imagined system, and to confront these solutions with reality, that is to say, to evaluate alternative solutions in relation to criteria of feasibility, efficiency, operability and environmental, societal or dimensional constraints. Going further towards achievement naturally involves risks. If success is to be achieved, it is better to analyse and evaluate these risks as soon as the first engineering data are available, and then evaluate them throughout the development process for each technical decision.

If the decision is made to realize one of the alternative solutions, it is also necessary to verify that the imagined solution (or theoretical solution) satisfies the expected characteristics (requirements). At the end of development, it must finally validate that the realized solution (product, service or organization) corresponds to the initial need. Given the necessary number of transformations between the perceived need and its realization (see the cycle of needs - Chapter 5.2.2, Volume 2, Systems Opportunities and Requirements [Faisandier2 2013]), there is strong presumption that there is no total adequacy. The only way to arrive at an adequate solution is to proceed in successive steps and iterations in each step, while checking that the outputs produced by each activity correspond to inputs and expectations (see development cycle - Chapter 6.1.2, Volume 1, System Notion and Engineering of Systems [Faisandier1 2015]).

If this talk seems only to be good sense, in fact it is not. In a natural and innate way, every human being is absolutely not in this approach of evaluating alternative solutions, verifying, justifying and validating. Knowledge and know-how must be acquired to ensure that the result is achieved in all cases. These are the knowledge and practices that this volume addresses.

> This volume explains, details, justifies the concepts, approaches and their implementation to evaluate and prove a system.

The engineering of a system is not just about posing the problem or identifying an opportunity, defining requirements, and defining an architectural and design solution using models and descriptions. Of course, these activities are essential and basic in order to qualify for the first maturity level of engineering (see Categories of processes and maps - Chapter 5.4.2, Volume 1, System Notion and Engineering of Systems [Faisandier1 2015]).

Maturity level 1 does not guarantee the systematic achievement of an engineering result, even by applying standard processes. To obtain a satisfactory engineering result, it is necessary to master the various transformations to pass from the need to the conceived solution, justifying each engineering data. This mastery takes part in the second maturity level of engineering. It is obtained by inserting tasks of evaluation, verification and validation of the handled entities (requirements, functions, system elements, interfaces, architectures, design, etc.) into the basic activities. This allows for realistic requirements, an architectural and design solution that is efficient with respect to the applicable requirements, and relevant to the operational need.

The same talk applies to the other aspect of development, i.e. the integration of the system (product, service, organization or combination of these elements). Integration does not only consist of assembling the implemented system elements that make up the system, and in making tests of good operation. This assembly and testing are the basic activities of the first maturity level. To obtain an implemented and complete system that satisfies the expected functionalities and operations, the assigned objectives and the specified constraints, it is necessary to verify and validate, step by step, that these have been taken into account in the realization and integration, according to an established strategy. This strategy is essential knowing that in a complex system it is impossible to verify and validate all combinations of events, scenarios and potential situations in the life of the future system, in a limited time. The definition of verification and validation actions, the establishment and execution of the verification and validation strategy, take place at maturity level 2 of integration.

Content and structure of this book

The aim of this book is to help industry and tertiary professionals in the definition, evaluation, verification, validation and justification of the properties of any system. This book presents the principles, concepts, activities and methods to:

- evaluate the properties of a defined or conceived (but not implemented) system, performing the activities of the System Analysis Process;

- verify and validate these properties throughout the development (i.e. on the definition of the system and its realization as a product, service or organization); this is done, via the verification and validation processes, by comparing the properties of the system with one or more requirements repositories;

- verify and validate applicable requirements that are sources of the system properties;

- integrate the implemented system elements (components), in order to obtain a ready-to-use product, service or organization.

It successively presents:

- ◆ A definition of the main terms used in systems engineering, including evaluation, verification, validation and integration aspects, in Chapter 2

- ◆ Reminders of notions explained in Volume 1, System Notion and Engineering of Systems [Faisandier1 2015], as well as extensions of the fundamentals of system development, in Chapter 3, which includes:
    - the notion of system
    - the life cycle processes for operational concepts definition and system definition
    - the life cycle processes for the integration of the system
    - ontology elements for the system definition and integration
    - the iterative and recursive approach of processes
    - the principles of intellectual creation, and the sources of errors introduced by architects, designers and producers

- the justification for the implementation of the evaluation, verification and validation approaches considering the errors mentioned above
- The assessment of system properties (system analysis) in Chapter 4, which includes:
  - concepts and principles relating to the assessment of properties, such as performance, efficiency, costs, environmental and technological constraints, technical risks; trade-off analysis
  - description of the assessment process activities
  - corresponding ontology elements
  - practical implementation
- The verification and validation of the system in Chapter 5, which includes:
  - concepts and approaches, such as verification and validation actions, verification and validation techniques or methods, verification and validation strategy, justification or verification matrix
  - description of the verification process activities
  - description of the validation process activities
  - corresponding ontology elements
  - practical implementation
- The integration of the implemented system elements in Chapter 6, which includes:
  - concepts and principles relating to the integration, such as integration methods, integration strategy, interfacing with other system life cycle processes
  - enabling system notion applied to the system integration
  - description of the integration process activities
  - corresponding ontology elements
  - practical implementation
- Some useful and recent references in accordance with this series of books in Chapter 7
- Annexes, which provide templates for plans, justification matrix, integration procedure, verification and validation procedures in Chapter 8

2 TERMINOLOGY

2.1 Terms and definitions

Terms and their definition(s) provided below (in alphabetical order) are used in Systems Engineering matters.

TERM	DEFINITION
(System) Architecting	Architecting is the deliberate manipulation of structure to achieve desired system behaviour and properties. (MIT-ESD)
Architectural Characteristic	An Architectural Characteristic concerns a System Architecture (essentially Physical Architecture, but sometimes Logical Architecture). It is the result of an intended arrangement of its composing elements that addresses particular System Life Cycle Abilities. In other words, Architectural Characteristics endow the system with specific life-cycle stages abilities (for example: operability, maintainability, reliability, execution efficiency, learnability, etc.).
(System) Architecture	The fundamental organization of a system embodied in its constituent parts, their relationships to each other, and to the environment, and the principles guiding its design and evolution. The architecture associated with a System of Interest is conceptual, and is realized through an architectural description. (IEEE 1471) Fundamental concepts or properties of a system in its environment embodied in its elements, relationships, and in the principles of its design and evolution. (ISO/IEC 42010) System architecture is the embodiment of concepts, and the allocation of physical / informational function to elements of form, and definition of interfaces among the elements and with the surrounding context. (Ed. Crawley - MIT) A system **architecture** is defined through several views, typically a logical view (including functional, behavioural, temporal models), a physical view and other optional views
Acquirer	The stakeholder that acquires or procures a product or a service from a supplier. (ISO/IEC 15288)
Behavioural Architecture	This term is the contraction of "Behavioural view of the system Architecture". An arrangement of functions and their sub-functions and interfaces (internal and external) which defines the execution sequencing, conditions for control or data-flow and the performance requirements to satisfy the requirements baseline. (Adapted from ISO/IEC 26702). A behavioural architecture can be described as a set of inter-related scenarios.
Constraint	(1) A restriction, limit, or regulation imposed on a product, project, or process. (2) A type of requirement or design feature that cannot be traded off. (ANSI/EIA 632)

Chapter 2. Terminology and abbreviations

TERM	DEFINITION
(System) Design	(System) Design includes activities to conceive a system that answers an intended purpose, using principles and concepts; it includes assessments and decisions to select elements that compose the system, fit the architecture of the system, and comply traded-off System Requirements. (System) Design is the complete set of detailed models, properties or characteristics described into a form suitable for implementation.
Design Property	A Design Property is a (quantitative) measure or estimate of an Architectural Characteristic associated with a System Element, a Physical Interface, a Physical Architecture. It is a property obtained by defining the elements through assignment of non-functional requirements, or obtained through estimate, analysis, study, calculation, simulation of a specific / particular aspect, or obtained by definition in the case of an existing System Element. If the defined System Element complies with a (set of) requirement(s), the design property(ies) should "equal" the requirement(s). Otherwise one has to identify the difference or non-conformance which treatment could conclude to modify the requirement(s), or the design, or identify a deviation.
Development (stage)	The life-cycle stage that includes concept definition, system definition, system realization, and transverse activities. Concept definition activity includes mission and/or business opportunity analysis and stakeholder requirements definition; system definition activity includes stakeholder requirements definition, system requirements definition, logical and physical architecture definition, design, and trade-off analysis activities. Realization includes implementation, and integration activities. Transverse activities include verification, validation and (project) management activities.
Engineering	(1) The application of scientific knowledge to practical problems, or the creation of useful things. The traditional fields of mechanical engineering, electrical engineering, etc. are included in this definition. [Checkland P.B. 1999] (2) To (cleverly) arrange for something to happen. [Checkland P.B. 1999]
Engineering Entity	Engineering meta-data. Can be a set of Stakeholder Requirements, a set of System Requirements, a Scenario (for logical architecture), a Physical Architecture, a Component (System element), a Physical Interface, etc.
Function	An action, a task, or an activity performed to achieve a desired outcome. A function uses input flows and generates output flows.
Functional Architecture	This term is the contraction of "Functional view of the system Architecture". A set of functions and their sub-functions that defines the transformations of input flows into output flows performed by the system to achieve its mission.
Functional Interface	A functional interface is constituted of an input flow or an output flow between two functions so that they may exchange material, energy, and/or information.

Chapter 2. Terminology and abbreviations

TERM	DEFINITION
Implemented Element	An implemented element is a System Element that has been implemented. In the case of hardware it is marked with a part / serial number.
Input-output Flow	An input or an output, or a trigger of a function (used for input-output flow, control flow). An Input-output Flow can be made of material and/or energy, and/or information.
Interface	An interface includes generally both functional and physical elements. Functional and physical interfaces are distinct.
Logical Architecture	This term is the contraction of "Logical view of the system Architecture". The set of Functional view / Architecture, Behavioural view / Architecture and Temporal view / Architecture of a system.
Measured Property	A Measured Property is a characteristic of the Implemented Element established after its implementation. The Measured Properties characterize the Implemented Element when it is completely realized, verified and validated. If the Implemented Element complies with a Design Property, the Measured Property should equal the Design Property. Otherwise one has to identify the difference or non-conformance which treatment could conclude to modify the Design and possibly the related Requirements, or to modify (correct, repair) the Implemented Element, or to identify a deviation.
Need	Necessity or desire expected by an "end user", expressed in terms of a client, of an end user; service, objective, capability expected from the future system by the end users. Equivalent to expectation; includes user requirements. A need could not be formally expressed or drafted.
Operational Concept	A general description of the way in which a product or service or enterprise (organisation) is used or operates.
Operational Mode	An operational situation / configuration of a system characterized by its active functions. Inside an operational mode the system can perform specific operational scenarios, and so activate the corresponding functions. (Same as state)
Operational Scenario	A set of actions or functions representing the dynamic of exchanges between the functions allowing the system to achieve a mission or a service.
Physical Architecture	This term is the contraction of "Physical view of the system Architecture". An arrangement of physical elements, which provides the designed solution for a product, a service, an organization or an enterprise, intended to satisfy the requirements of the logical architecture and the requirement baseline. (Adapted from ISO/IEC 26702 = ISO/IEC 24748-4)
Physical Interface	A physical interface is a System Element that binds physically two System Elements.

TERM	DEFINITION
Port	Element of a System Element that allows binding this System Element to another one with a Physical Interface.
Rationale or Justification	Argument that provides the justification or relevance for the selection of an engineering element (e.g., requirement, function, flow, system element, interface, physical architecture, etc.).
Requirement	Statement that identifies an expected characteristic of a product, service, enterprise or process operational, functional, or constraint, which is unambiguous, testable or measurable, and necessary for product, service, enterprise or process acceptability. (Adapted from ISO/IEC 26702 = ISO/IEC 24748-4)
Requirements Engineering	Subset of systems engineering concerned with discovering, developing, analysing, validating, communicating and managing requirements. (ISO/IEC 29148)
Risk	(Used for technical risk in engineering) An event having a probability of occurrence and a gravity degree on its consequence onto the system mission or on other characteristics. A risk is a measure of the combination of vulnerability and of hazard or threat.
Stakeholder	Party having a right, share or claim in a system or in its possession of characteristics that meets that party's needs and expectations (ISO/IEC 15288:2015). N.B.: Stakeholders include, but are not limited to, end users, end user organizations, supporters, developers, producers, trainers, maintainers, disposers, acquirers, customers, operators, supplier organizations and regulatory bodies. (ISO/IEC 29148)
Stakeholder Requirement	Necessity or desire expected by an end user, formally drafted and expressed in terms of a client, of an end user; service, objective, capability expected from the future system by the end users. Equivalent to expectation; includes user requirements.
Supplier	An organization or an individual that enters into an agreement with the acquirer for the supply of a product or a service. (ISO/IEC 15288)
System Element (component)	(1) A member of a set of elements that constitutes a system. A system element is a discrete part of a system that can be implemented to fulfil design properties. A system element can be hardware, software, data, humans, processes (e.g., processes for providing service to users), procedures (e.g., operator instructions), facilities, materials, and naturally occurring entities (e.g., water, organisms, minerals), or any combination. (Adapted from ISO/IEC 15288:2008) (2) Generic term used for System of Interest, system, System Element. A physical element (user or operator role, hardware, software) that composes a system. A system element may be seen in its turn as a system (i.e. sub-system) and be engineered in a system block at lower level.

Chapter 2. Terminology and abbreviations

TERM	DEFINITION
System Element Requirement	A requirement applicable to the System Element defined during the architecture and design of the system that contains it. Its nature is similar to a Stakeholder Requirement; the Stakeholder is the system that contains this System Element. It is elaborated from the system physical and logical architectures elements and from the System Requirements of the system.
System(s) Engineering	The application of systems science (principles, rules, laws) through a set of defined activities to answer complex practical problems or opportunities creating, arranging concepts and implementing them.
System of Systems	(1) Two or more systems that are separately defined but operate together to perform a common goal. [Checkland P.B. 1999] (2) An assemblage of components which individually may be regarded as systems, and which possess two additional properties: - Operational Independence of the Components: if the system-of-systems is disassembled into its component systems the component systems must be able to usefully operate independently; that is, the components fulfil customer-operator purposes on their own. - Managerial Independence of the Components: the component systems not only can operate dependently, they do operate independently; the component systems are separately acquired and integrated but maintain a continuing operational existence independent of the system-of-systems. [Maier M. 2009]
System Requirement	(1) Requirement applicable to a system; see requirement. (2) A statement that converts a stakeholder requirement into technical, usable terms for architectural and design activities using a semantic code (natural language, mathematical express, arithmetic, geometric, or software language, etc.).
Temporal Architecture	This term is the contraction of "Temporal view of the system Architecture". A temporal classification of the functions of the system depending on their frequency level of execution. The temporal architecture includes definition of synchronous and asynchronous aspects of functions. The decision monitoring inside a system follows the same temporal classification as decisions are related to monitoring of functions.
Transition-of-Mode	A transition of mode characterizes the transition (change of state) between two successive Operational Modes of a system.
User (end user)	Individual who or group that benefits from a system during its utilization. (ISO/IEC 15288)

Chapter 2. Terminology and abbreviations

Specific terms and their definition(s) relating to evaluation, verification, validation and integration are provided below, in alphabetical order.

TERM	DEFINITION
Assessment Criterion	Characteristic used to assess and compare engineering elements between them.
Assessment Score	Calculated value using a combination of weighted criteria with each other and the marks assigned for each criterion to an assessed engineering element.
Assessment Selection	The element that is the object of assessment. Tree root in a multi-criteria model.
Cost	Estimated, assessed or calculated value of the amount allocated to an object (e.g. Component, Physical Architecture, Physical interface), or an activity of the life cycle, expressed in a given currency at a given time.
Error	An error is a deviation between the expected result and the obtained result (production of a non-compliant result or unavailability of an expected result when a fault is solicited); the error can manifest as a failure.
Failure	A failure or breakdown is an abnormal behaviour of the system (product, service, organization) by production of a non-compliant result or not producing the result, observable by a user of the system. It is the concrete manifestation of an error.
Fault / Defect	A fault is a design or implementation flaw, or a degradation of one or more system elements that can lead to an error if it is solicited. This is the primary cause of an error; error that can cause a system failure.
Implemented system element	A system element that has been realised (fabricated, assembled); it is identified by a serial number. Examples: software application, mechanical component, electrical component, operator role, procedure, protocol, etc.
Integration aggregate	Composition of several "implemented system elements" of the system which are assembled, on which a set of verification and/or validation actions is performed.
Justification / Rationale	A justification action is to demonstrate the relevance of an action (choice, decision, approach, etc.) by providing evidence or written arguments.
Mistake	A mistake is done by a designer / implementer and produces the existence of a fault / defect in a product, or even a vulnerability.
Qualification	The qualification (of the definition) is a formal contractual act attesting the conformity of the definition regarding the system and/or stakeholder requirements, and the reproducibility of this definition.

Chapter 2. Terminology and abbreviations

TERM	DEFINITION
Validation	It is the set of activities intended to acquire confidence in the ability of a system to accomplish its mission, under specified conditions of use. These activities aim to establish the compliance of the system with respect to its technical requirements and/or to its stakeholder requirements. Validation is a transverse (or support) process of the life cycle of a system.
(Final) Validation	Final validation is an instance of the validation process applied to the overall system at the end of development.
Validation Action	A validation action describes what must be validated and how: the object submitted to validation, the expected result of the execution of this action, the method or technique to be applied, at which system decomposition level the action must be executed.
Verification	It is the set of activities intended to determine whether the system meets the expected design properties or characteristics. These may be general or specific. These activities aim to look for defects in the system. Verification is a transverse (or support) process of the life cycle of a system.
Verification Action	A verification action describes what must be checked and how: the object submitted to verification, the expected result of the execution of this action, the method or technique to be applied, at which system decomposition level the action must be executed.
Vulnerability	Characteristic or fault / defect of the system that can be exploited by a threat or an internal or external hazard and can generate a failure.

2.2 Abbreviations

ACRONYM	DEFINITION
COTS	Component On The Shelf
eFFBD	Enhanced Functional Flow Block Diagram
IBD	Internal Block Diagram (SysML)
IES	Integration Enabling System
INCOSE	International Council On Systems Engineering
I/O	Input - Output
PBD	Physical Block Diagram
SBS	System Breakdown Structure
SE	Systems Engineering
SEBoK	Systems Engineering Body of Knowledge
SOA	Service Oriented Architecture
SoI / SOI	System-of-Interest
SysML	System Modelling Language (OMG)

3 REMINDERS AND EXTENSIONS ABOUT FUNDAMENTALS FOR THE DEVELOPMENT OF THE SYSTEM

This chapter presents a reminder of some of the fundamentals explained in Volume 1 of this series of books, System Notion and Engineering of Systems [Faisandier1 2015], as well as the necessary extensions that must be known in order to properly develop any system (i.e. to engineer and integrate) and master, via the evaluation, verification and validation approaches, the errors potentially introduced.

This chapter deals with the following subjects:

- The notion of system - section 3.1
- The life cycle processes to define the operational concepts and the system - section 3.2
- The life cycle processes to develop the system - section 3.3
- The ontology elements for engineering the system - section 3.4
- The intellectual creation mechanism, and sources of errors - section 3.5

Chapter 3.1. Reminders and extensions about fundamentals: System notion

3.1 Notion of System

This chapter provides the selected definition of *system* as outlined and discussed in Volume 1 of this series, System Notion and Engineering of Systems [Faisandier1 2015], the main views for defining any system, and the consequences of adopting this definition.

Selected definition of *system*

The development of any complex product, service or organization is facilitated when viewed as a system, and when the systems engineering discipline is used. There are several definitions of the notion of system, but the following is most useful for carrying out the engineering. It is this definition that is selected in the volumes of this series of books:

> A system is characterised by a set of **components** (technological equipment, software, human operators, materials, procedures, services); the components **interact** strongly and exchange **flows** of matter, energy and information in a given **context** or environment. This set satisfies **needs**, expectations, requirements; it achieves a **mission** associated with prescribed **objectives** that allow answering a **purpose**.

This concise definition is relatively complete. It covers all multidisciplinary systems presently listed, at least the systems subject to concrete realisations that answer the technological, organizational or strategic needs of human societies.

Figure 1 graphically shows a model of the selected definition of the *system* entity.

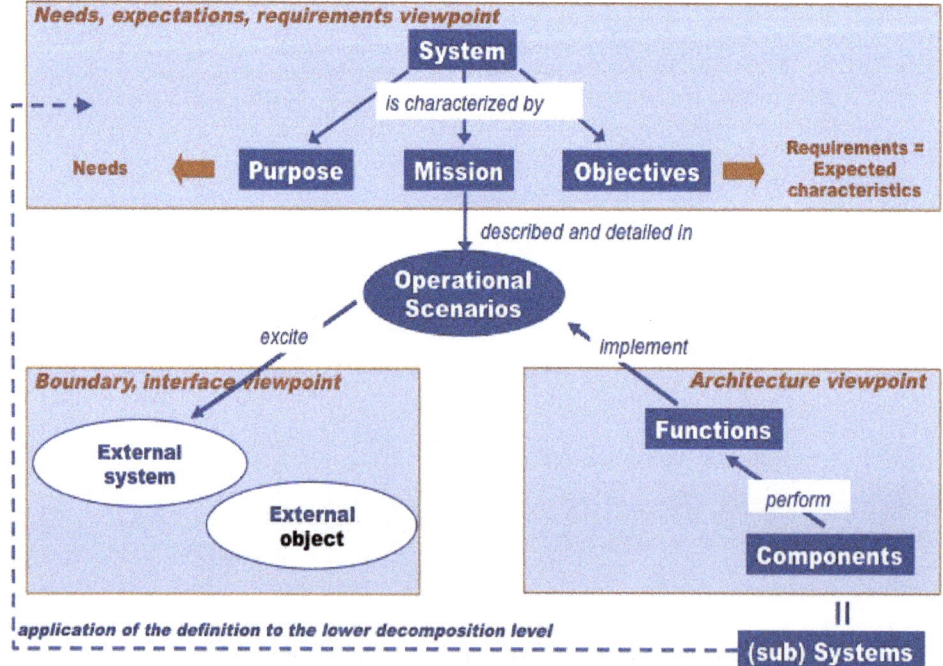

Figure 1 - A model of the definition of system

The reading of this figure is the following:

- ♦ A system is characterized by a purpose, a mission and objectives. These synthetic items are detailed in terms of needs and expectations, which are refined as technical requirements, that is to say as expected characteristics (prescriptive).

- ♦ The mission is detailed and described in the form of operational scenarios. The operational scenarios excite objects and systems that are external to the system-of-interest, and vice versa, to exchange matter, energy and/or information. These exchanges enable the identification of interfaces and interactions, and so the functional and physical boundaries of the system.

- ♦ The operational scenarios are implemented through actions (or functions) grouped as logical architecture. The functions or actions are performed by concrete components that compose a physical architecture.

- ♦ When the number of concrete components is important (more than about ten), it is recommended to form (sub) systems. Each (sub) system that composes the system-of-interest is a system; therefore, it is characterized by the same generic items as the parent system, but differently instantiated. This is the way one can obtain a hierarchical composition of the system-of-interest.

Three major viewpoints

To characterize any kind of system, which one wants to engineer, we consider the three following viewpoints:

Needs and Requirements viewpoint - The following characteristics allow a global understanding of the system in its context of use; the system is seen as a black box.

- **Purpose** - The purpose expresses the relevance of the system in its context of use. It presents the final aim of the system in its environment. The expression of the purpose is obtained considering the system from outside, and answering questions such as "Why would this system exist? What should be the relevance of its existence in the given context? To what aim would this future system participate? What would be the usage of the system?"

- **Mission** - The mission expresses the set of services the system has to deliver; this means the main or global transformation, which the system performs. The expression of the mission is obtained answering questions, such as "What does the system perform? What is it supposed to do? What does it transform? What service does it deliver?"

- **Objectives** - The objectives are the items that quantify the mission and the services of the system as measurable data or properties related to space, time or effectiveness. The expression of the objectives is obtained answering the questions "How many input items does the system transform into output items? With which effectiveness does the system perform the transformations, the services? Within what duration? Within which frequency? Etc."

- **Operational concept and scenario(s)** - The expression of the mission is generally an abstraction of a set of actions performed by the system in its context of use and in defined operational conditions. This set of actions constitutes operational scenarios that describe how these actions are linked to each other, as well as the matter, energy and/or information they exchange with the objects or systems of the context.

During the engineering, all the items above are defined, analysed and refined iteratively, until a global and complete understanding of the system in its context, through needs then stakeholder requirements, is obtained. The system requirements are defined translating, detailing and refining the stakeholder requirements. The definition of needs and requirements is described and detailed in Volume 2 of this series, Systems Opportunities and Requirements [Faisandier2 2013].

Architecture viewpoint - Two complementary views of the system compose the fundamental basis of the system existence:

- The **logical architecture**, (contraction of "logical view of the architecture of the system") which is a structure or composition of **functions** related to:
  - transformations of matter, energy and/or information (**input-output flow**), which are represented through one or several functional architecture model(s); the **functional architecture** includes the exchanges of input-output flows between the functions of the system, and also with the external elements to the system;
  - dynamics of operational **scenarios** that enable the system to execute the expected services in all the situations, as identified in the life cycle of the system; these situations are generally named **operational modes**; the operational modes and scenarios compose one or several **behavioural architecture** model(s); the behavioural architecture can be represented as scenarios of transition of modes, and/or as scenarios of functions; in this last case, the functions are linked with logical constructs that show how the functions of the system have to be executed in order to achieve the mission;
  - the temporal aspects, which enable figuring out the execution frequency levels of the functions, the synchronisms and asynchronisms; these aspects are represented through one or several **temporal architecture** model(s).
- The **physical architecture**, (contraction of "physical view of the architecture of the system") which is a structure or composition of concrete **components** that perform the functions of the system, and of their physical links or connections, named **physical interfaces**, that carry the input-output flows (matter, energy and/or information) between and outside of the functions. These components can be tangible or intangible (systems or components, such as equipment or products made of technologies, software and/or human roles).

The logical architecture and the physical architecture correspond. The interactions between components are defined by interfaces whose complexity strongly depends on the way these two architectures are composed. The architecture aspects are detailed in the Volume 3 of this series, Systems Architecture and Design [Faisandier3 2013].

Boundary and interface viewpoint - The boundary of the system depends on what the architect of the system includes or not in the system-of-interest in terms of functionalities and concrete components. This viewpoint is deduced from the previous one above; it focuses on the exchanges between the components of the system and with the objects or systems external to the system-of-interest; one distinguishes:

- The **functional interfaces,** which are the **input-output flows** exchanged between the functions of the system and with the functions of the external objects or systems. These flows are made of matter, energy and/or information. They allow describing the dynamic interactions. They are essential items to define the interoperability of the system with the outside world.
- The **physical interfaces**, which are the physical connections between the concrete components of the system, and with the external objects. The physical interfaces carry the input-output flows. Their nature depends on technologies that are used by the connected components; a component may be considered as a physical interface; examples of physical interfaces: electrical cable, fluid pipe, signal carrier, magnetic field, protocol for exchange, procedure.

An interface is a complex notion, but its usage is essential for the system architecture definition. The choices for allocation of functions to components can create more or less complicated interfacing and so generate more or less simple architectures.

Recursion of the definition of a system - decomposition

As indicated in the model of the selected definition of the system notion - see Figure 1, in its turn a component of the physical architecture can be seen as a system. In a common sense, one talks more of a sub-system in order to characterize the hierarchical level difference between the system-of-interest and the systems composing it.

Nevertheless, to be able to utilize the term "sub-system", one must keep the coherence of the definition. That is to say, a sub-system has firstly to be a system such as defined previously. A sub-system is supposed to be characterized by its purpose within the system that includes it, its own mission, its own objectives, its architecture of components, its boundaries and interfaces with the other sub-systems of the encompassing system. This means that the definition of the system notion we selected is **recursive**.

In the perspective of a recursive definition of the system notion, the systems can be refined in layers of systems, which gives birth to the System Breakdown Structure notion - SBS, used to control the engineering of systems including numerous components. Figure 2 portrays a generic example of the hierarchical breakdown of systems. The decomposition into systems stops when the component is no more considered as a system; that is the case of the technological components, named "end-product", in this figure.

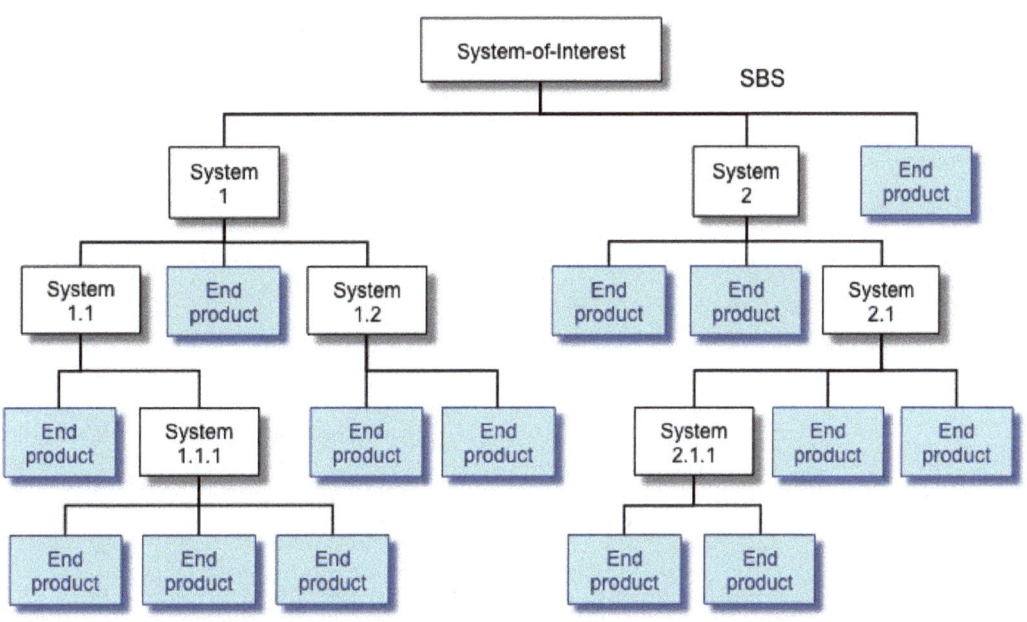

Figure 2 - Hierarchical breakdown of the system-of-interest (SBS)

Such a hierarchical breakdown expresses a parent-child relationship, which is in fact restrictive for several reasons, and whose utilization often generates misunderstandings depending on the interpretation of this representation. The parent-child relationship assumes that an element has only one parent; this assumption bothers an architect who desires to study the various architecture alternatives of a single father-system. Some engineering tools do not provide such flexibility. This hierarchical representation no longer takes into account the exchanges between the systems or components, nor the dynamic of exchanges. Rather than using the term "hierarchical breakdown levels", we prefer to use the term "systems encapsulation", like Russian dolls.

Chapter 3.1. Reminders and extensions about fundamentals: System notion

The encapsulation relationship has the advantage of being able to locate the component in its context and its connections to the other components or systems - see Figure 3.

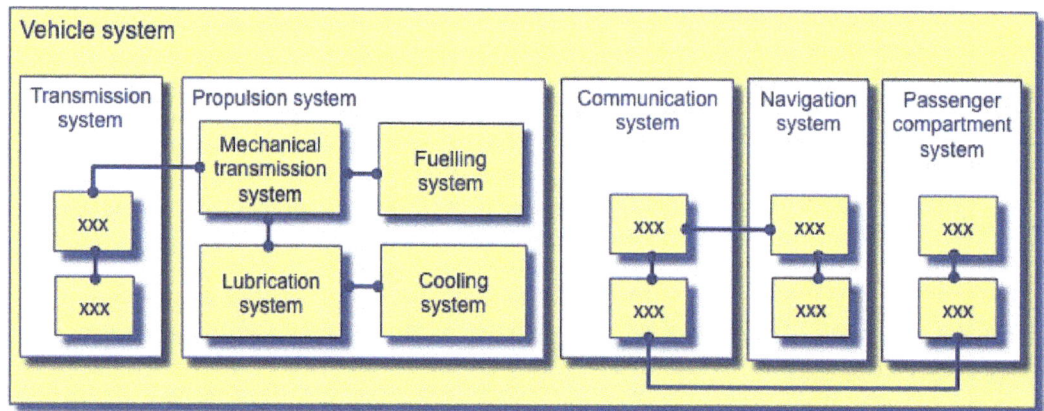

Figure 3 - Encapsulation of systems with their connections

The encapsulation notion can also be used with the system-of-interest, of higher hierarchical level, which is immersed in its context, assuming that this context can be considered as a system. Figure 4 illustrates this aspect.

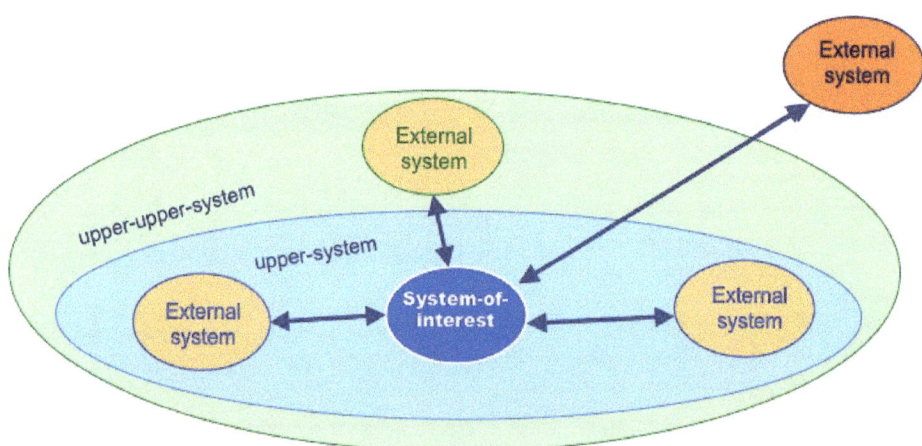

Figure 4 - Encapsulation of the system-of-interest in its context

Chapter 3.1. Reminders and extensions about fundamentals: System notion

Conclusion: A single engineering method suitable for all

The definition of system is recursive, and its engineering is also recursive. The engineering consists of defining each characteristic of a system and to refine them.

At each abstraction level and for each system, the same activities and the same modelling techniques are necessary, with possibly few adaptations - see Figure 5:

- An N+1 level (sub) system, with defined architecture and design, is a solution that answers a need expressed by the system of level N; to express the need of a (sub) system at level N+1, it is necessary to have defined the architecture and design of the level N system.

- The N level system, with defined architecture and design, is a solution that answers a need expressed by the (upper) N-1 level system; to express the need of the level N system, it is necessary to have defined the architecture and design of the level N-1 (upper) system.

In other words, it is not possible to define the needs for a system, if one has not yet modelled the architecture of the upper system (i.e., the context of use of the system).

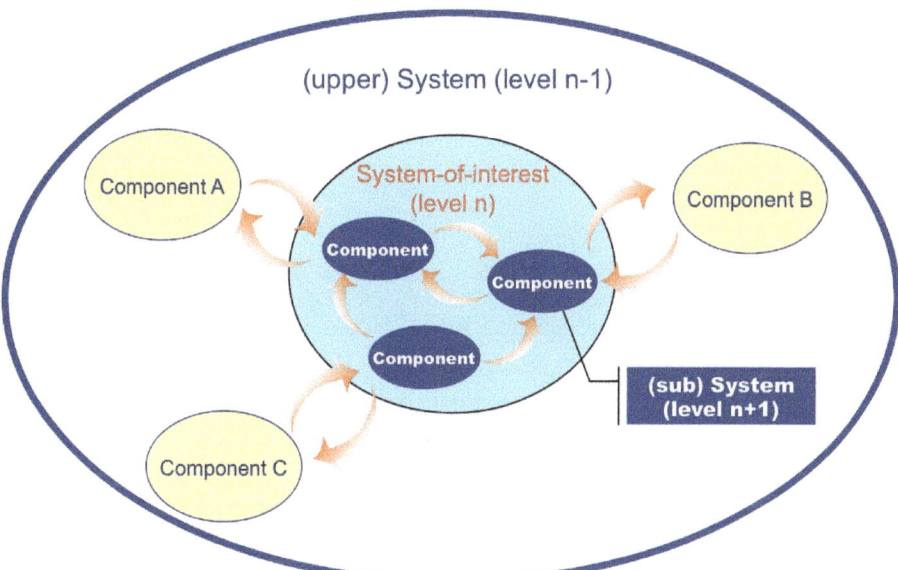

Figure 5 - Recursion of the definition of systems

At the organization stage, this means that acquirers and suppliers perform system architecture and design activities, on their own level, onto different systems.

3.2 Processes to define the concepts and the system

The selected definition of the system notion is recursive. It is used at each level of the decomposition of the system-of-interest into (sub) systems; it is instantiated for each (sub) system of this decomposition. The consequence is that the definition processes for any system are also recursive. That is to say, the same activities and tasks are executed at each level of the decomposition, but instantiated to the considered (sub) system.

Notion of process (reminder)

A process transforms inputs into outputs by adding value. The transformation is about material, energy, information or a combination of these. The transformation is performed on the form, in the space or time using means.

A process consists of federated activities around a **theme** in order to realise a product or a service. It uses methods, techniques, practices, and procedures to perform the required transformations. For example, an acquisition process transforms a need into a supply.

Initialization of the System Definition

The activities related to the Definition of a System are preceded by the activities related to the Definition of operational or technical Concepts, on which are based the solution(s) with respect to the concerned opportunity or problem. This is not THE solution at this stage of the definition, but food for thought about the ideas of principles on which potential solutions could be based.

Concepts Definition is the set of upstream activities that initialize the development, identifying issues, problems, improvements and opportunities, and/or needs to create a new system, or to improve an existing one, or to update a system. This part of engineering recursively uses the two following processes:

- **Mission or business analysis** process,
- **Stakeholder needs and requirements definition** process.

Outcomes of execution of the first process are used as inputs of the second process, which provides feedback to the first one. The next execution of these two processes can be parallelized - see upper side of Figure 6.

System Definition uses the following processes:

- **Stakeholder needs and requirements definition** process
- **System requirements definition** process
- **Logical architecture definition** process
- **Physical architecture definition** process

Iteration of System Definition processes

As previously, system definition processes are iteratively and overall concurrently executed. That is to say that each execution of a process improves its own engineering data, and provides feedback to other processes, which then enable the improvement of their data, and so on.

Chapter 3.2. Reminders and extensions about fundamentals: Processes for the definition

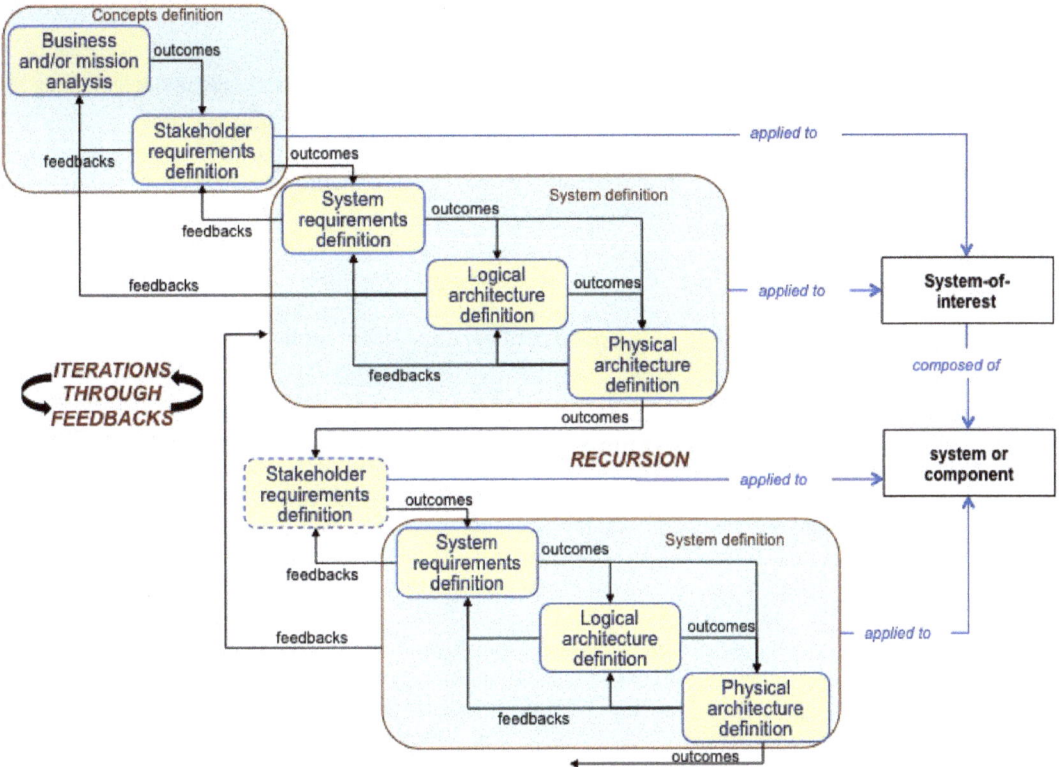

Figure 6 - Iteration and recursion of definition processes

A first iteration of these four processes can be done linearly, in the order cited above, to obtain the main system engineering data without seeking further details. This is obviously a first draft. Other drafts will be needed to refine the definition of engineering data. It is then preferable, and quite natural, to perform the following iterations by parallelizing the execution of processes.

Remark

Proceeding iteratively, gradually completing engineering data to move towards completeness, providing details and information by successive refinements, has the effect of working on the full scope of the system. It is also a way to apply the system vision. The consequence is that results are obtained quickly, all points are identified and analysed, and the engineering data form a coherent and relevant set at every moment.

This holistic approach is different from working with trees, such as a tree of needs, system requirements tree, functions tree, or components tree. Generally, a tree is obtained by a simple relationship (even simplistic) of parent-child decomposition; this relationship is one of the poorest in the engineering ontology. From experience, the approach with trees does not sufficiently highlight the complex interactions and interdependencies between internal manipulated elements, or exchanged with elements outside the system, early enough. The necessary coherence is made a posteriori, which often requires breaking what has been previously developed, resulting in a loss of time and unnecessary effort consumption. Achieving consistency of engineering data and their relevance is something difficult, but it is the major factor in the success of a system development project.

Recursion of System Definition processes

The system-of-interest physical architecture gives rise to concrete components of the system. These components (so called System Elements) may be either (sub) system, or technological components / end-products - see illustration Figure 2.

In the case of (sub) systems, the four processes mentioned above are instantiated and used for the Definition of each (sub) system in the same way as before. It should be noted that the stakeholder needs and requirements definition process is represented with a dotted line in Figure 6. This means that its use is not systematic:

- In the upper levels of abstraction, the degree of freedom, with respect to the solution, is relatively large; that is to say, the notion of need provides flexibility in decisions and choices (notion of real, perceived, retained, expressed need - see Volume 2, Systems Opportunities and Requirements [Faisandier2 2013]). It is appropriate to use this process to identify needs precisely.

- On the contrary, in the lower levels of abstraction (technological level), the degree of freedom is very low, because more natural laws of physics, for example, are applied. The process is not useful, and one can go directly to the development of technical requirements for System Elements (components).

Exit from System Definition

In the case where a System Element (component) of the system-of-interest is a technological end-product (no longer a decomposable component), the approach is different, because one exits the area of system engineering to enter into the area of technologies engineering. However, one must make the transition between these two areas - see Figure 7.

Implementation of System Element is the set of development activities that perform this transition. It uses the three following processes:

- **System Element requirements definition** process
- **System Element design definition** process
- **System Element implementation** process

The System Element requirements definition process is the same as the system requirements definition process, but applied to each concerned technological component (System Element). The other two processes, design definition and implementation, are specific to used technology/ies for the concerned System Element.

Technological components may consist of mechanical and/or electrical, electronic, hydraulic, chemical, biological, software components, operator roles, etc.

Chapter 3.2. Reminders and extensions about fundamentals: Processes for the definition

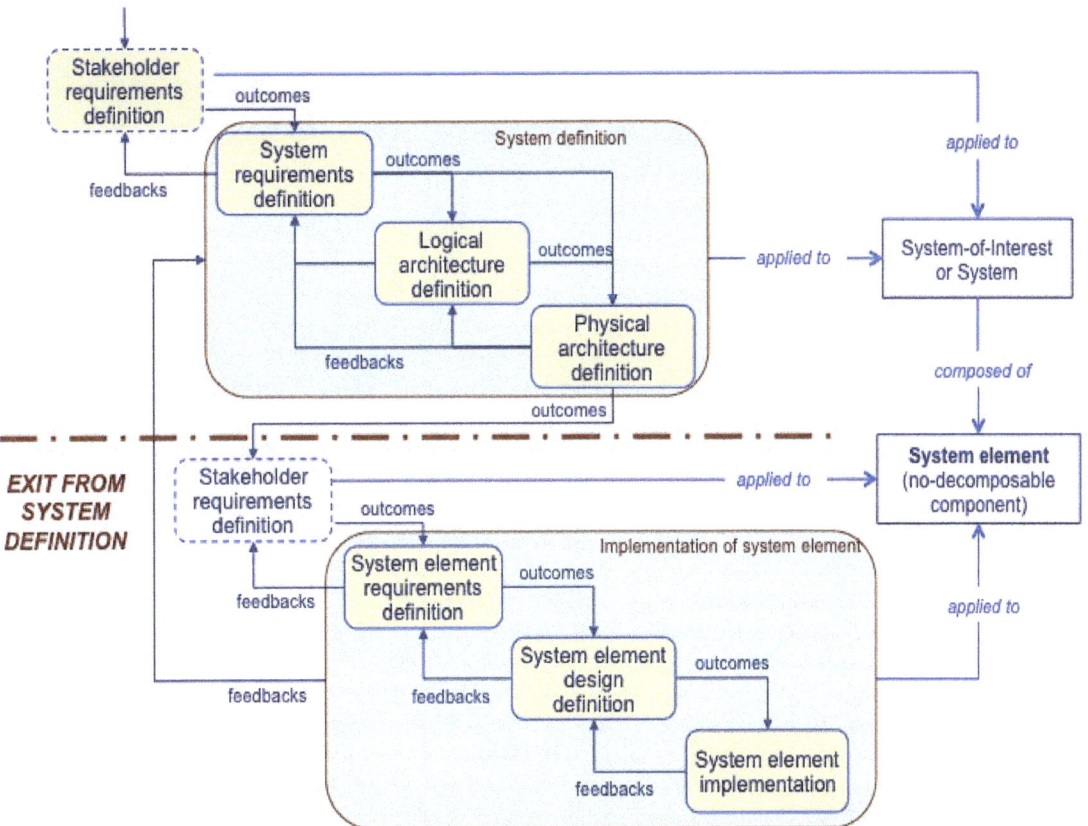

Figure 7 - Transition towards System Element implementation processes

3.3 Processes for developing the system

The present chapter introduces the notion of system-block - section 3.3.1, then describes:

- The development approach using processes and system-blocks - section 3.3.2
- The evaluation approach - section 3.3.3
- The verification and validation approach - section 3.3.4

3.3.1 Notion of system-block

To facilitate the development of a system composed of several elementary components (System Elements), the system is decomposed into levels of finer components and (sub) systems. So, each level comprises of only a small number of components and/or (sub) systems. This arrangement allows better control of complexity, in particular the complexity due to the interfaces and interactions between the components. For development of such systems, at each level and each component or (sub) system of the physical tree, a **system-block** is associated. Instead of having one large project for the global system-of-interest, one can have as many small projects as system-blocks, or gather within the same project several small system-blocks.

For each system-block there are corresponding definition activities and integration activities. For an essentially didactic and intended purpose, we present these activities in a V structure - see Figure 8.

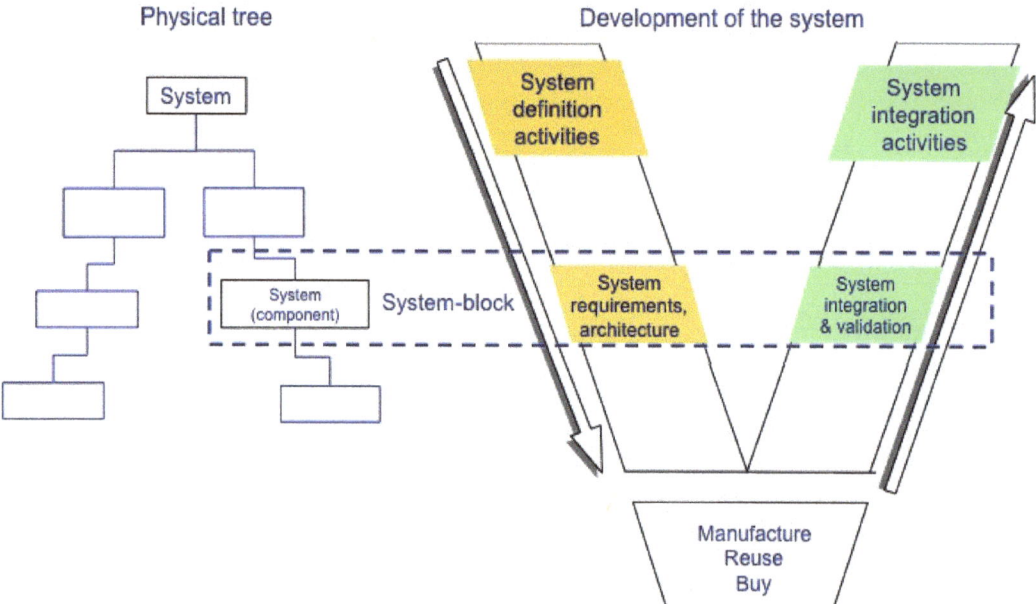

Figure 8 - Notion of system-block for development purpose

Chapter 3.3. Reminders and extensions about fundamentals: Processes for the development

3.3.2 Development approach using processes and system-blocks

Here, development activities being relatively numerous, are divided into three groups of processes, from the development viewpoint: basic technical and definition support processes, implementation processes and integration processes. To simplify the explanations, we take a top-down approach to development.

These three groups of processes are presented by a system-block level in a V structure - see Figure 9. In this figure, other processes were added, related to the system operational life, but which are not part of the development (upper right of the figure).

- The **definition** processes are placed on the descending branch of the V, on the left.
    - ♦ In the block of the context (i.e., the context or environment of the future system), the following two basic processes are distinguished:
        - the **mission or business analysis** process,
        - the **stakeholder needs and requirements definition** process.

    These two processes provide mechanisms for analysing the context of use and initializing the system development with associated requirements. As mentioned in the previous chapter, they are executed iteratively.

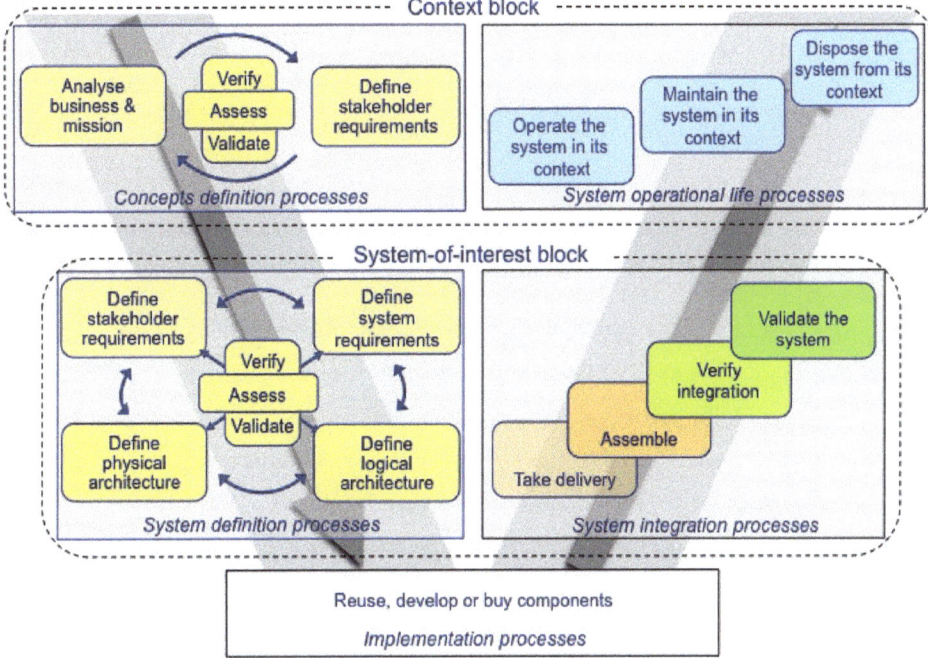

Figure 9 - System development processes

Note: In the context block, on the right-hand side of the V structure, the other processes are added related to the operational system life, even though they are not part of the development. The following processes are distinguished:

- the **system operation** process,
- the **system maintenance** process,
- the **system disposal** process.

35

Chapter 3.3. Reminders and extensions about fundamentals: Processes for the development

- ◆ In the system-of-interest block, and in the blocks of each system of each decomposition level, the four technical basic processes are distinguished:
    - the **stakeholder needs and requirements definition** process,
    - the **system requirements definition** process,
    - the **logical architecture definition** process,
    - the **physical architecture definition** process.

    These four processes provide the mechanisms to develop the system definition, in a controlled manner. As mentioned in the previous chapter, they are executed iteratively and recursively for each system, at each level of decomposition. This enables one to progressively generate a coherent set of engineering data, for each system, in particular stakeholder requirements, system requirements, architecture logical entities and physical entities architecture. In this block, the stakeholder needs and requirements process is used to supplement and consolidate these elements from the execution of the same process in the context block.

    In both blocks (context block and system-of-interest block), three definition support processes are associated with the technical basic processes:
    - the **system properties assessment** process,
    - the **verification** process,
    - the **validation** process.

    These three processes are used to make coherent and relevant all engineering data developed by the technical basic processes. They are called upon, when necessary, by each of the other processes.

- • The system components **implementation** processes are located at the bottom of the V structure. Figure 10 details the implementation processes group. The following processes are distinguished:
    - the **system element requirements definition** process,
    - the **system element design definition** process,
    - the **system element implementation** process,
    - the **system element properties assessment** process,
    - the **(system element) verification** process,
    - the **(system element) validation** process.

They are used either to manufacture the components defined in the latter system level, or to reuse components having been developed in other contexts, or to buy components from a catalogue. In the case of purchase, only a part of the activities of some of these processes is used. These processes do not concern the system level; they are at the level of technologies, and each technology has its own techniques and methods of engineering and implementation.

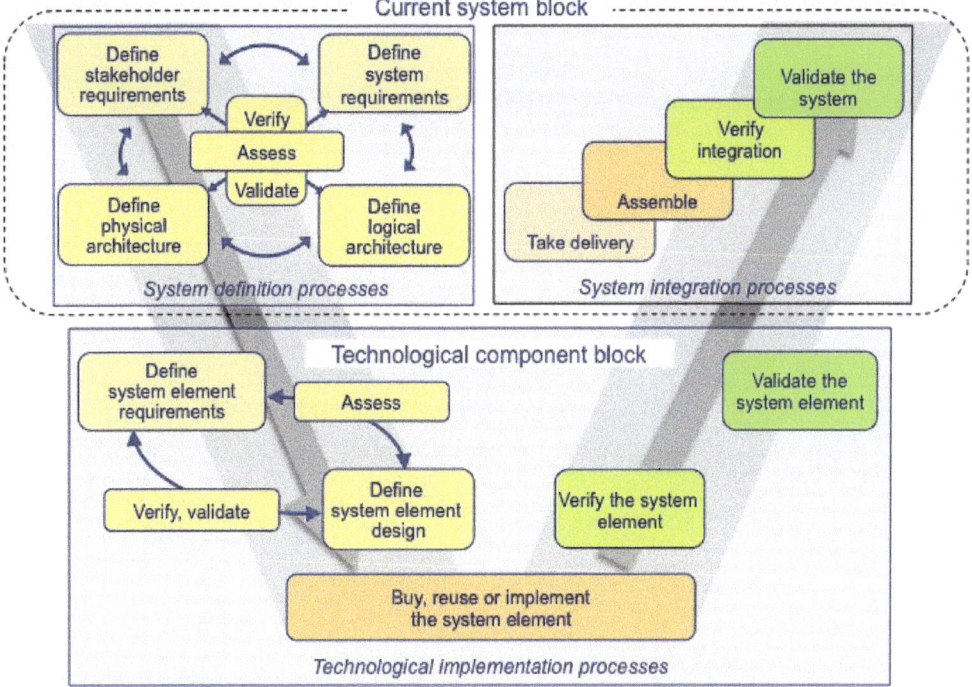

Figure 10 - System Element implementation processes

- In the system-of-interest block, and in the blocks of each system of each decomposition level, the system **integration** processes are placed on the ascending branch of the V, on the right. The following processes are distinguished:
    - the **implemented system elements reception** process,
    - the **implemented system elements assembly** process,
    - the **(system) verification** process,
    - the **(system) validation** process.

The integration processes are executed as linearly as possible; that is to say, the verification and validation processes should only find that the assembly solution properly satisfied architectural properties and system requirements, and was not used to tweak and tune. The effort of development can be reduced by making partial concrete mock-ups or simulations during definition activities, in order to characterize the properties with tests and experiments.

Methodological remarks

The V structure (sometimes called life cycle V model) presents the activities of the operational life of the system and the integration activities (right-hand side) mirroring the concepts definition activities and system definition activities (left-hand side). This representation means that "downstream" activities are only the consequence or result of "upstream" activities; in other words, that one integrates and uses the system, which was previously thought out and defined.

Without going into the details of the contents of each process, we can say that, if during the operational life of the system one does not encounter major difficulty, the purpose, mission, objectives, operational concept and operational scenarios have been well identified and described upstream, they have served effectively in the system definition, i.e., they have been correctly translated into requirements, and these requirements have been effectively used to identify a logical architecture, and the latter has been correctly projected onto a physical architecture, etc.

The reverse is quite true and verified in the field during actual projects. In this case, development teams spend their time during integration to solve problems by coring, and thus consume resources to fix the system rather than to innovate new systems. At worst, it is the users who spend their time in operation to look for a bypass to operating difficulties encountered, because of lacks in engineering (to gain time during development)!

3.3.3 System properties assessment approach

In the previous paragraphs, we mentioned the system properties assessment process as part of the development approaches. This process is considered as a development support process.

In a dynamic view, this process is used by other processes of concepts definition and system definition. Figure 11 presents its relationships with the technical basic processes.

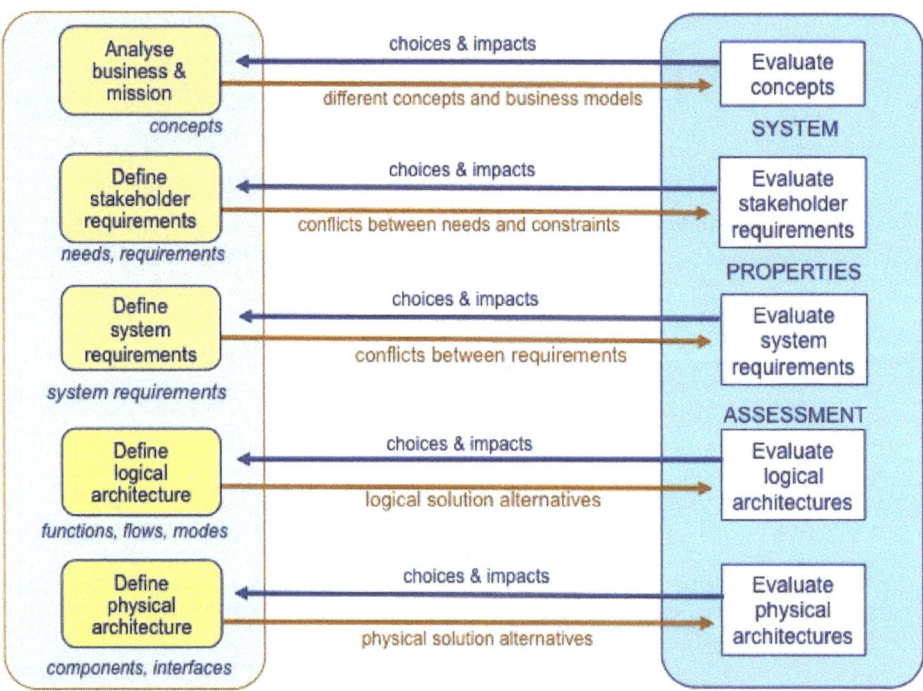

Figure 11 - Assessment process and approach

The system properties assessment process (called System Analysis process in standards and handbooks) provides a rigorous approach to make technical decisions, whether to:

- solve conflicts and requirements choices,
- perform the analysis of different properties using methods, techniques and tools (simulations, calculations, algorithms, specific models, etc.), or
- evaluate and compare physical and logical alternative solutions using objective criteria and impact analysis.

Chapter 3.3. Reminders and extensions about fundamentals: Processes for the development

3.3.4 Verification and validation approach

Similarly, the verification and validation processes are an integral part of development approaches. They are definition support processes and also integration support processes. They do not concern the same objects in both cases; they are instantiated according to the items to verify and validate.

These two generic processes are actually transverse for all the activities and processes of the entire system life cycle. It is in this sense that one considers verification and validation approaches. These approaches are defined through a **verification and validation strategy**, which covers all activities of the life cycle, in a consistent and relevant way according to the identified objectives and constraints.

Figure 12 shows that every technical process calls for verification and validation processes. Each activity of a process, and each output of this activity can (or should) be subject to verification actions and validation actions.

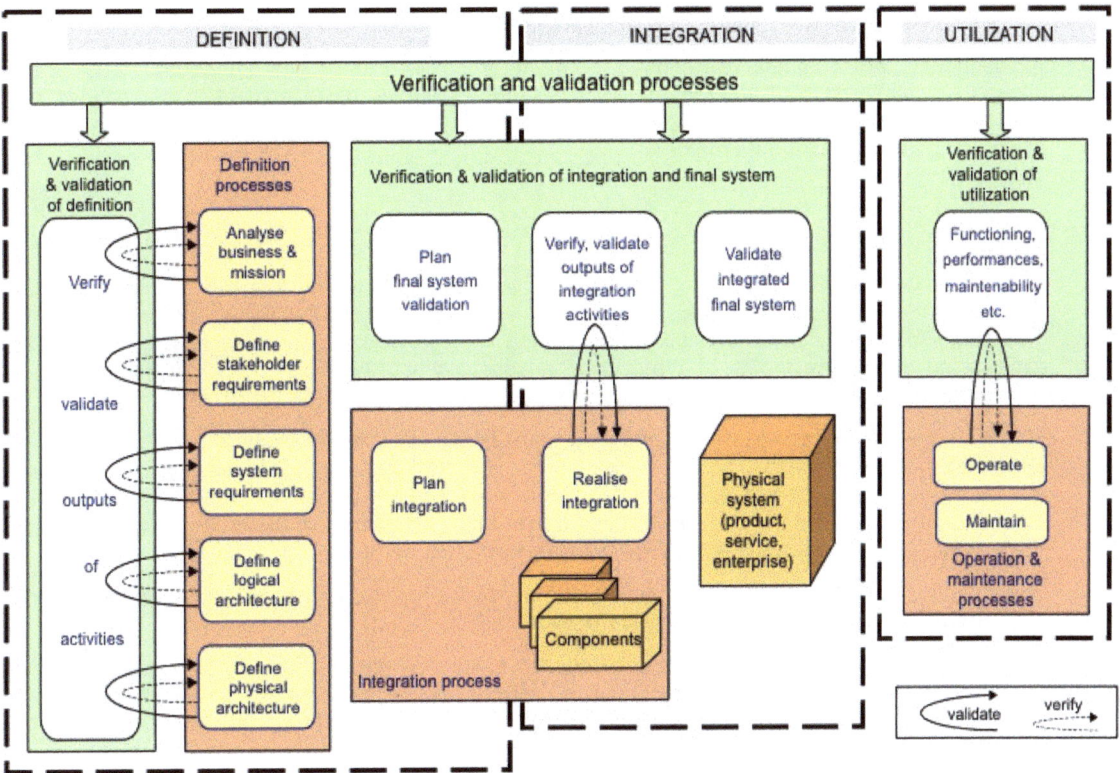

Figure 12 - Verification and validation processes and approach

3.4 Ontology for engineering a system

The ontology for engineering systems is reduced to an engineering generic meta-data model here. Such a model consists of **entities**, each representing a concept, or a basic notion used in one or more processes, and potential **relationships** between these entities within a single process or between processes.

A meta-data model (briefly meta-model) is defined from the process activities. It is a view that focuses on the data used (input data) and produced (output data) by transformation activities of processes. In the present case, it is a generic engineering meta-model. The suggested model represents a kind of skeleton of the whole engineering processes; a skeleton, because the meta-model is restricted to only data used, produced and exchanged. The proposed generic engineering meta-model is a part of the **language / ontology** of systems engineering.

Each of the next two sections gives a view of the engineering meta-model:

- section 3.4.1 gives a simplified overview for development of a system, and explains its use in two development approaches, top-down and bottom-up,
- section 3.4.2 gives a simplified overview for system definition.

Subsequently, the synthetic description of each process includes the concerned part of the engineering meta-model. Each volume in this series describes more accurately the entities and relationships corresponding to covered topics. Finally, Volume 7 contains the complete system engineering generic meta-data model.

3.4.1 Ontology for development of the system

A general, but simplified view, of the generic engineering meta-model part dedicated to the development of the system is shown in Figure 13. The rectangular boxes represent the engineering meta-data; the arrows are the relationships between these meta-data, which allow, in particular, establishing traceability tables or matrices. Relationships are essential, because they provide a **consistent** set of engineering data for a given system.

Throughout its development, a system is apprehended with the outcomes of processes and engineering activities; mainly stakeholder needs and requirements definition process, system requirements definition process, architecture processes, design definition process, implementation and integration processes. At any given time, these outcomes represent the technical maturity state of the system; the states are briefly called "**as required**", "**as designed**", and "**as realized**".

The comparison between these different results or states enables one to control the convergence, step by step, toward a solution that meets the needs and expectations (requirements).

Each state "as xy" is established using models and assertions that must be consistent with those of the preceding "as xx" state. Comparisons and different findings between the system "as required", the system "as designed" and the system "as realised" provide the means to control the technical development.

The entities and relationships in Figure 13 support different development approaches and related life cycle activities, such as top-down, reverse engineering / bottom-up, or mixed approaches.

- A top-down development approach can be supported as follows, starting from the left of the diagram:
    - The **Stakeholders** write **Stakeholder Requirements** that are then translated by **System Requirements**.
    - **System Requirements** are then satisfied by (logical definition part) or taken into account (physical definition part) architecture and design entities, in particular **Functions,** which are

Chapter 3.4. Reminders and extensions about fundamentals: Ontology for engineering a system

performed by **Components**, and **Input / Output Flows**, which are carried by **Physical interfaces**.

♦ **Components** and **Physical interfaces** are implemented by **Implemented Components** according to selected technology (ies).

♦ Finally, these last elements are integrated in intermediate **Aggregates** until the integrated system-of-interest (product, service, organisation / enterprise).

In definition and design activities, **Design Properties** and architectural characteristics of the system can be assessed and compared to **Assessment Criteria** (extracted or defined from System Requirements). This enables one to identify potential problems, and so to tune the design, or to decide on modification of requirements.

When the system is realised and in use / operation, **Measured Properties** can be compared to **Design Properties** and to **System Requirements**. These comparisons are used to verify the correct operation of the system in use with respect to what was required or what was conceived and designed.

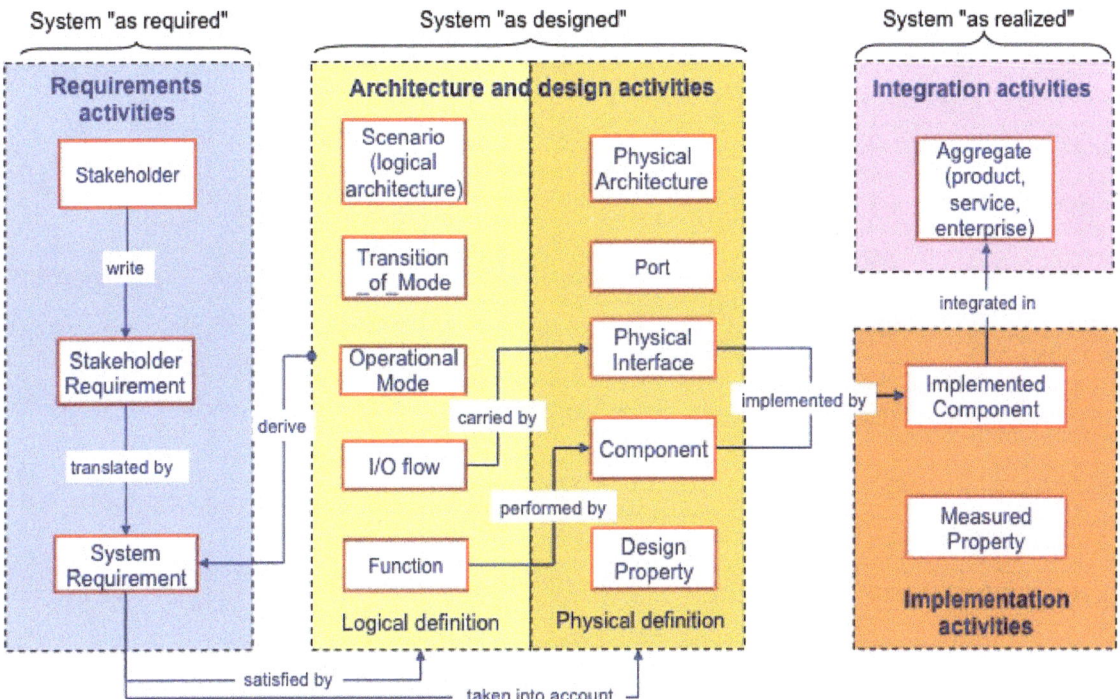

Figure 13 - Simplified view of the engineering meta-model for development

- A bottom-up approach, associated with reverse engineering, can also be supported by this meta-model, in order to restructure or re-architect existing systems to meet their changing context of use. We start this time on the right of the diagram as follows:

    ♦ In step one, the **Implemented Components** are identified during a physical analysis of the product, service, or organisation. Then the **Components** and corresponding **Physical interfaces** are identified in turn.

- In step two, reverse engineering consists of identifying the **Functions** performed by the **Components**, and the **Input / Output Flows** carried by the **Physical interfaces**. This makes possible the development of physical architecture and functional architecture models of the existing system.

- Finally, it is possible to characterize the existing system with a set of **Design Properties** and **System Requirements** extracted from the physical and functional architecture models. Additional modelling activities are often necessary to characterize the behavioural and temporal aspects (logical architecture).

- Any changes to the existing system are then possible in a top-down approach (reengineering), considering new **System Requirements** related to evolutions, etc.

3.4.2 Ontology for the definition of the system

Figure 14 presents a simplified overview of the generic engineering meta-model dedicated to concepts definition and system definition processes and activities. As mentioned earlier, these meta-model elements form a skeleton of activities and processes.

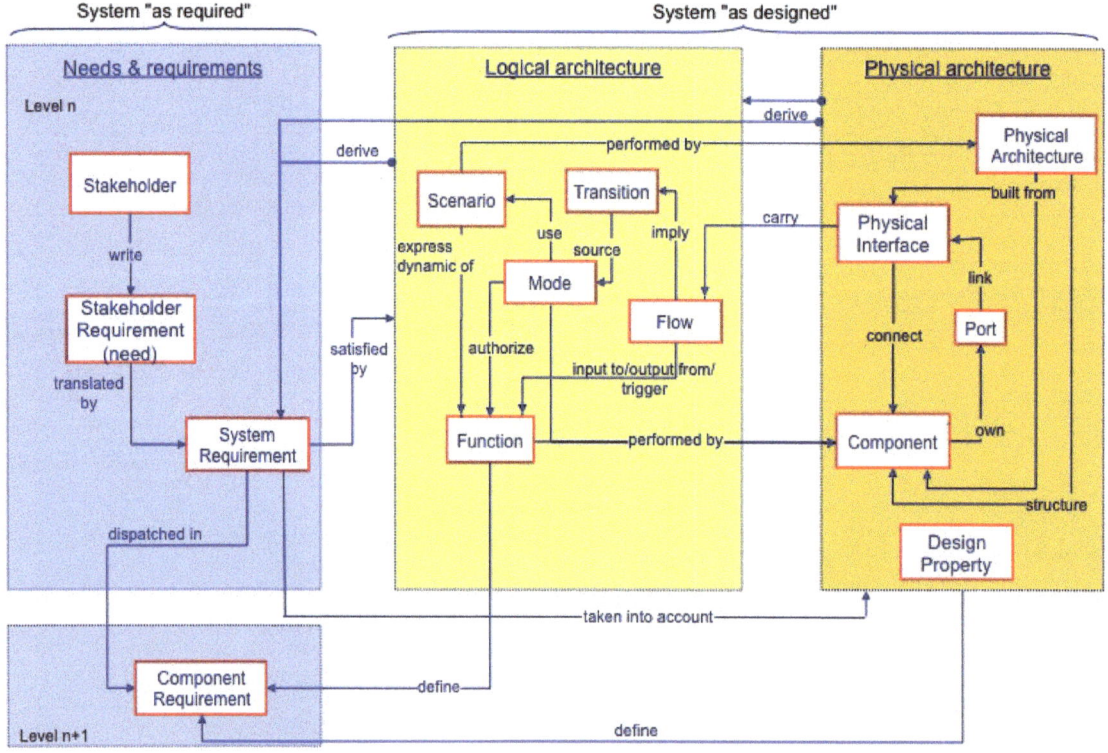

Figure 14 - Simplified view of the engineering meta-model for definition

Chapter 3.4. Reminders and extensions about fundamentals: Ontology for engineering a system

During the execution of these processes and the use of associated modelling techniques, experts in systems engineering:

- Establish all necessary instances of these meta-data (boxes).

- Provide information attributes of these instances with specific values. The generic attributes represent the meta-data characteristics, and the specific values for an instance particularises this instance among others.

- Instantiate the generic relationships between the meta-data to establish traceability links between data.

Implemented by software applications such as Databases, or Repository, or tools like Spreadsheet, these relationships can generate traceability matrices. These matrices are useful for a variety of actions, such as verifications, audits, impact analysis (in case of changes), justification, technical risks management, verification of data consistency between them, etc.

Notes

In the previous two figures and in the figures of the remainder of this volume representing parts of meta-model, the relationships between two entities are mentioned in one direction only, such as "a **Need / Stakeholder Requirement** is translated by a **System Requirement**". The inverse relationship is still possible, but it is not mentioned in the figures, for example "a **System Requirement** translates a **Need / Stakeholder Requirement**".

In Figure 13 and Figure 14 **Need** and **Stakeholder Requirement** are equivalent names; the same for **Mode** and **Operational Mode**, as well as **Transition** and **Transition of modes**, **Flow** and **Input / Output Flow**.

In Figure 14, the entity **Component Requirement** concerns the **Stakeholder Requirements** applicable to a **Component** of the system-of-interest architecture; the instances of this entity enable one to make the link with level n+1 of the system tree. They can be exported to a new database. This arrangement avoids putting in a single database the system-of-interest data and the data of its components therefore, avoiding databases with monstrous information volumes.

Cardinalities (maximum and minimum number of possible entities in the relationship) are not mentioned in this volume; Volume 7 indicates them.

Finally, in the previous two sections, there is little or no mention of entities and relationships related to system property assessment activities and verification, validation and integration activities. These ontology elements are described in detail in the following chapters.

3.5 Intellectual creation and origin of errors

The intellectual creation mechanism is at the core of all engineering activities (whatever the type of engineering), because engineering is the domain of ideas and of their transformation.

It should be noted that our academic education uses more analytical methods (for example, analysis of what exists, problems, texts, etc.) than holistic methods (for example, expressing a problem, creation of generic models, creation of concepts, etc.). The acquisition of knowledge obviously goes through the analysis of what exists, the details of things, the detailed understanding of phenomena, rules and laws. But, whatever the discipline, it must or should include the understanding of the whole, the relationships between and dependencies with phenomena and laws, the knowledge of principles that govern phenomena and laws, as well as the relationships between the disciplines. For example, the knowledge and utilization of the causal law (cause - effect relationship) that is essential in engineering.

The present chapter describes:

- A mental behavioural model - section 3.5.1
- The intellectual creation mechanism and its application to engineering activities - section 3.5.2
- The relevance of assessment, verification and validation approaches with respect to human "reliability" - section 3.5.3

3.5.1 A mental behavioural model

Certain psychologists tried to understand the acquisition, memorization, and restitution phenomena of knowledge; essential elements of intellectual creation. Figure 15 portrays some elements of a possible model coming from studies by George Armitage Miller [Miller 1956].

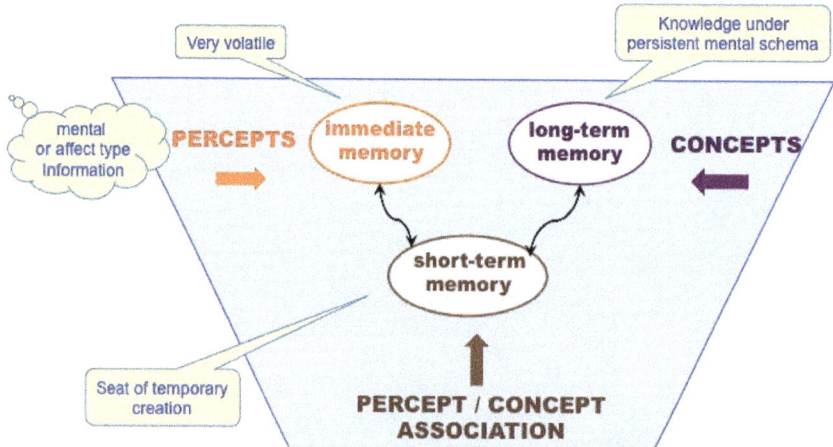

Figure 15 - A knowledge acquisition model

The model considers three types of memory:

- The **immediate memory** that records **perceptions** (percepts) (i.e. words, images, symbols, smells, feeling, etc.). This memory is very volatile; that is to say that the arrival of a new percept may exclude the previous percept recorded in this memory, because of a lack or interruption of attention.

- The **long-term memory** that contains **concepts** (ideas, knowledge) recorded in the form of persistent mental schema.

- The **short-term memory** that enables **percept-concept associations**, that is to say to establish relationships between what is perceived with what is known. This memory is the seat of temporary creation, where the idea can be emitted on the basis of the percept-concept association.

Perceptions (percepts) concern objects and sensations related to situations or events in the environment of the individual, which he captures through his five senses (sight, hearing, smell, taste, touch).

Concepts are acquired and encrusted knowledge during the life of the individual, through teaching, learning lived experience, both consciously and unconsciously.

If an individual perceives something that has no relationship with recorded concepts, the association is not possible. Either the percept is ignored, and there cannot be a creation; or it is temporarily memorized, and there is a transient creation; or it is memorized as a new concept, and there is a persistent creation. The acquisition of new concepts or ideas is generally the fact of the percept-concept association repetitions for incrustation, from which come the practice of exercises.

One will easily verify this model with any kind of learning, for a language, a technology, a method, an intellectual or manual technique.

This model may also help us to understand the learning, creation difficulties, and also the potential errors that may be made during the process:

- erroneous perception,
- concept repatriation difficulties: memory defect, so no recognition,
- percept-concept association defect: non-creation, erroneous interpretation, non-realistic creation, etc.

The difficulties encountered, and the errors produced during the mechanism of intellectual creation are amplified if the individual receives, at the same time, information of a mental or affect type unrelated to the subject treated (noises). Any diversion disrupts this process, whether the information is negative or positive.

3.5.2 Creation mechanism and systems engineering activities

Intellectual creation or conceptualization mechanism

The previous model may serve as a basis to explain how our intellect functions during the engineering works. An attentive observation of the way we proceed shows that we intellectually work with more or less short **analysis-synthesis** cycles: we perceive and analyse, for example, objects, ideas and their relationships, then we try to formulate, express and represent syntheses through association with yet incrusted concepts. Figure 16 summarizes this mechanism:

Chapter 3.5. Reminders and extensions about fundamentals: Intellectual creation and origin of errors

- As a general manner, the **analysis** work contains two major tasks: the perception and the understanding.
    - ◆ The **perception** consists of, as we have seen, identifying words, drawings, objects, sounds, etc.
    - ◆ The **understanding** consists of associating perceptions with concepts recorded in our memory, sorting or discriminating concepts to do something with, or not. The analysis work ends here.
- The **synthesis** work also contains two major tasks: imagination and expression.
    - ◆ The **imagination** consists of transforming the concepts into mental images, ideas, then organizing them, structuring between them, globalizing, detailing and clarifying them.
    - ◆ The **expression** consists of transcribing the ideas and mental images into words, drawings, symbols, sounds, etc., using known standards of the concerned community.

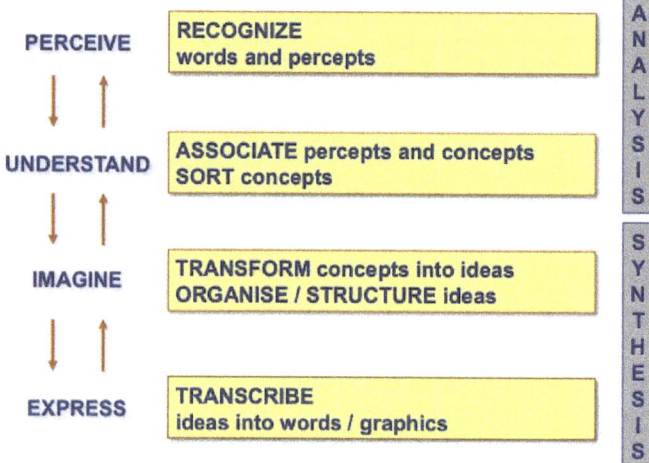

Figure 16 - Intellectual creation / conceptualization mechanism

Our intellect does not operate in a sequential manner; that is to say it does not execute only one linear sequence, "Perceive, Understand, Imagine, Express". It goes back and forth several times between these actions, which is perfect regarding the mechanism result:

- Between Perception and Understanding, it checks the right perception and the right relationship between the percept and the recorded concept. So, it prevents losses of attention or wrong interpretations. As a consequence, omissions and inappropriateness are avoided.
- Between imagination and understanding, it checks that the transformation of concepts into ideas is meaningful; that is to say that there is no deviation caused by obsession with one idea (blockage) or by anarchy of ideas (mental confusion).
- Between expression and imagination, it checks that the expression represents the original idea or image well, and finally that the addressees of the expression in turn are able to understand and utilize it.

Chapter 3.5. Reminders and extensions about fundamentals: Intellectual creation and origin of errors

Application to engineering activities

Each engineering activity or group of activities uses this analysis / synthesis cycle. Here are utilization examples within some systems engineering activities:

- During the **operational concepts and needs definition**, there is:
  - analysis of the context of use, i.e.:
    - perception of the objects of this context,
    - understanding of their relationships with the system-of-interest, then
  - synthesis through
    - imagination of services to be delivered,
    - expression of these services and expression of the system mission through globalization of services and exchanges.
- During the **system technical requirements definition**, there is:
  - analysis of the needs, i.e.:
    - perception reading the needs,
    - detailed understanding of ideas expressed, through refinement keeping the original idea, then
  - comparison with knowledge about the technical subject through synthesis; that is to say
    - imagination of characteristics to endow the system, then
    - expression of system technical requirements using a predefined (standardized) typology and expression rules.
- During the **logical architecture definition**, there is:
  - analysis of applicable system technical requirements, i.e.:
    - perception reading the system requirements,
    - understanding of the expected characteristics, recognition of corresponding logical entities (or concepts), such as functions, flows, modes, etc., then
  - synthesis through
    - imagination of functions architectures, using known or to be created patterns,
    - expression with functioning scenarios, state-transition diagrams, etc.
- During the **physical architecture definition**, there is:
  - analysis of the logical architecture through
    - perception of logical entities (functions, flows, modes, etc.),
    - understanding of functioning, and mapping logical entities with known components capable of performance and effectiveness, then
  - synthesis through
    - imagination of structures implementing these logical entities (functions, flows) through partitioning and allocation, respecting operational and environmental conditions as expressed in the system requirements, then
    - expression of the physical architecture with various diagrams, drawings, pictures, etc.

3.5.3 Relevance of evaluation, verification and validation approaches with respect to human "reliability"

Human reliability within the creation / conceptualization activity

The fundamental question in this volume is: Can a human being create or conceive something, such as a system (product, service or organization) without ever making a mistake? In the light of daily observations, the answer is obviously negative - *errare humanum est*. It must be admitted that error is an integral part of human thought and action, as well as the learning of every practice.

Section 3.5.1, that presents a model of mental behaviour, explicitly describes three brain activities (perception, concept recognition, association percept - concept) and potential mistakes that may be made during execution. The following table provides examples of potential human errors occurring in each of the four major activities of the creation / conceptualization mechanism presented in section 3.5.2, consequences on mental artefacts, and the consequences on the created object (product, service, organization).

Activity	Human error	Consequence on mental artefact	Consequence on product, service
Perception	Loss of attention	Loss of elements	Incomplete product, service
Understanding	Misinterpretation	Lack of organization, prioritization	Unsuitable product, service
Imagination	Fixation or anarchy	Unsuitable idea	Product, service not optimized
Expression	No respect of the rules of expression	Personal style or language	Incomprehensible, not standardized product, service

Studies in human reliability show that the human being, as operator, performs his physical or intellectual activities, within a given frame, with an error rate (number of errors / number of elementary actions performed) 10^{-3} if well concentrated and well trained, 10^{-2} in normal operation, 10^{-1} or less if disturbed. As an example, the production of errors by a designer in the realization of a software piece is on average 1 error for 200 lines of code, whatever the language used.

Human reliability is influenced by at least three factors:

- the **quantity** of information to be processed in a given time (which affects perceptions and associations percepts - concepts),
- the **complexity** of the activity executed, i.e., the number of operations to be performed and the number of interactions between the elementary operations,
- multiple environmental **disturbances** (noise, interruptions, stress, emotions, etc.).

Finally, in a systemic view of our societies, we see that social, economic and cultural contexts produce their own shortcomings, which generate more or less serious effects in products, services and organizations. These effects are generally referred to as "non-quality".

These findings argue in favour of the establishment of justification approaches by architects, designers and implementers. Not by mistrust, but by insurance of all the stakeholders, including for the comfort of the architects, designers and implementers. The justification consists of explaining why a particular solution, a property of design, a requirement, an action, an element, etc., has been decided.

Notion of justification

> A **justification** is to demonstrate the relevance of an action (choice, decision, approach, etc.) by providing evidence, findings, or arguments. The **evidence** comes from verification or validation actions. **Arguments** come from evaluations against criteria, or from expert opinion (analogy, feedback, etc.), or reasoning (logic, mathematics, probabilistic). **Relevance** is not understood in an absolute way, but in relation to a context (notion of pragmatics).

The **justification** is obtained in three ways, non-exclusive but often complementary:

- ◆ **evaluation** by comparison of candidate elements against pre-defined criteria (or characteristic), for the purpose of researching or selecting relevant elements in a given context or situation,
- ◆ **validation** by comparison of selected elements with respect to an operational reference, in order that their users accept these elements,
- ◆ **verification** by comparison of these elements with respect to a technical reference of know-how and rules, often called the state of the art, for the purpose of detecting defects.

Initial complex systems (products, services, organizations) created, conceived and realized by humans have more, or less defects, inabilities and vulnerabilities. "Initial system" means the first version of the system, or prototype; the subsequent versions or evolutions correct the defects and tend, in general, to improve the functioning of the system.

The justification approaches (evaluation, validation and verification) are normal and even indispensable with regard to the stakes, whether they are:

- technical order:
  - ◆ to research the adequacy or conformity of the product, service or organization with the needs of the stakeholders,
  - ◆ for ease of use, i.e., operability, by users,
- social order:
  - ◆ for safety of persons and goods,
  - ◆ for the quality of service provided to users,
- economical order:
  - ◆ with regard to industrial competition,
  - ◆ costs and time limits,
  - ◆ contracts for supplies and services.

These stakes must guide the various actions of justification to be deployed throughout the development (engineering, realization and integration).

Chapter 3.5. Reminders and extensions about fundamentals: Intellectual creation and origin of errors

Origin of system failures

Figure 17 illustrates the relationship between a *failure* of a system and its origins or causes. A failure, that is observable by the user of the system, is global and affects the normal operation of the system.

This failure results from an *error* that is produced by the execution or use of a system element. This error propagates through the system elements and may produce cascading errors until the failure of the system is observed. This propagation may be longer or shorter and may even remain latent for a long time; it depends, in particular, on the interdependence of the system elements and the frequency of execution of the system elements functionalities.

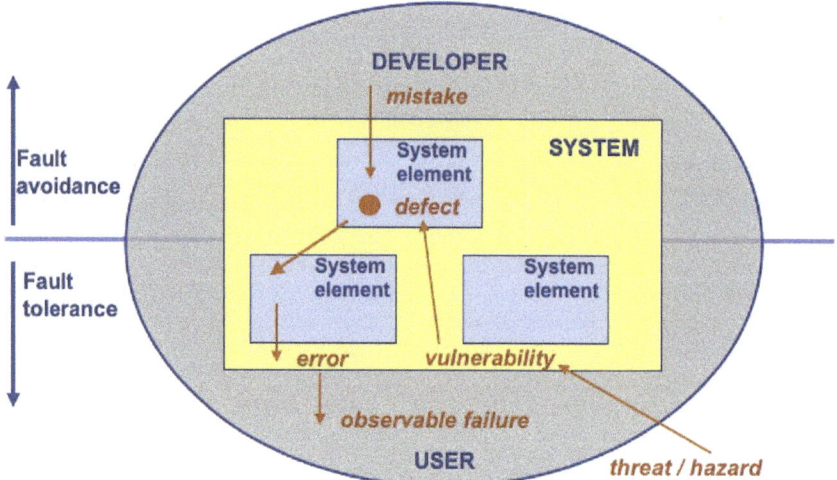

Figure 17 - Origin of observable failures

The error is due to a *defect* of the implemented system element. There are several types of defects depending on the nature or technology of the system element, and there may be several different causes for this defect; for example, the abnormal wear of a moving mechanical part, corrosion, poorly sized parameter, coding error, *vulnerability* to environmental conditions, potential *threats* or *hazards*, etc.

Nevertheless, all these causes of defects can be seen as "*mistakes*" committed by the architects, designers, and implementers during their tasks of creation, design, realization, or those of the decision-makers through their choices or decisions. These mistakes are not intentional, of course; they can be due to omissions, environmental changes, unforeseen events, inadequate sizing due to lack of knowledge of certain phenomena, etc., or even potential threats or hazards not apprehended.

Subsequently, in this book, we distinguish as often as possible the entities or concepts supported by the terms mistake, defect, error and failure; the adopted definitions follow.

Chapter 3.5. Reminders and extensions about fundamentals: Intellectual creation and origin of errors

> A **mistake** is committed by an architect, designer or implementer during the creation tasks, and gives rise to a defect in the system (product, service, organization) or even vulnerability.
>
> See Note 1.

Note 1: Mistake

The term "mistake" is used here, do not confuse it with the term "error", despite a connotation that is often culturally negative.

> A **defect** is an imperfection inherent in the product, service or organization, introduced during engineering or realization; the defect can lead to an error if it is solicited.
>
> See Note 2.

Note 2: Defect

The term "defect" covers internal defects in the product, service, organization that may be due to:

- a mistake made by the developer who has unintentionally introduced this defect for various reasons (insufficient knowledge of the studied subject or methods, inexperience, non-exhaustiveness of the study, incoherence, ignorance of environmental conditions, etc.)

- forgotten by the developer of the robustness to external or internal threats or hazards (which are normally the subject of a risk analysis).

> An **error** is a gap between the expected result when executing the functionality of a system (product, service, organization) and the result actually delivered by the system.
>
> See Note 3.

Note 3: Error

An error may not be observable (user-detectable). An error is typically functional. It can spread and give rise to other errors, which complicates the search for causes or origins.

The term "Undesired Event" (UE), often used in some dependability analysis techniques, is synonymous with error (elementary UE) and/or failure (head UE).

> A **failure** is an abnormal behaviour of the system (product, service, organization) by production of a non-compliant result or by lack of result, observable by a user of the system.
>
> This is the actual manifestation of an error.

> A **vulnerability** is a lack (or defect) of the robustness of the system that can be exploited by a threat or internal or external hazard and that may generate a failure.

To conclude the explanations in Figure 17, any complex system has defects and/or vulnerabilities, and these systems must live and/or operate in the presence of potential or proven defects; especially for systems with a long service life. The fundamental challenge is to control errors and failures.

What is possible by acting, at the same time, on causes and consequences as follows:

- On the causes, with **fault avoidance** development techniques - upper part of the figure. To avoid introducing defects, one must avoid making mistakes; this is achieved through mastery of activities and processes (training and learning) and quality control. Despite these provisions, defects are introduced into the system; the verification and validation activities contribute strongly to detecting them and eliminating them.

- On the consequences through **fault tolerance** development techniques - the lower part of the figure. In complex systems, despite the execution of verification and validation actions, residual defects and/or vulnerabilities remain. Fault tolerance techniques allow delivery of the service or expected results within specified limits in terms of availability, and/or safety, and/or survivability; for example, containment barriers, passive, active or semi-active redundancies or replications; these aspects are dealt with in Volume 5 of this series.

At any level of perception, all quality control approaches are aimed at controlling the creation of potential mistakes and defects, and controlling the propagation of errors and failures. The justification approaches contribute strongly to this control, either by systematic evaluation of certain properties of the system and their combinations (notably with regard to emerging properties), or by relevant verification and validation actions facing technical or programmatic risks.

In conclusion of this chapter, we will remember that the main origins of the defects of the systems, we conceive and realize, are essentially human, due to lack of experience, knowledge, method, lack of attention, concentration, verification, bad expression, uncontrolled imagination, no questioning of habits, by precipitation, etc.

The following chapters provide many elements for mastering the engineering and integration of the systems we create, imagine, and realize.

4 SYSTEM PROPERTIES ASSESSMENT (SYSTEM ANALYSIS)

This chapter presents the necessary set of means to carry out the assessment activities of the properties of the system during its definition. The assessment process provides the requesting processes with the elements of analysis of the system properties and justification arguments, so as to choose between different alternative solutions.

This chapter presents:

- The assessment concepts and principles - Chapter 4.1, which includes:
  - Trade-off studies - section 4.1.1
  - Cost analyses - section 4.1.2
  - Technical risk analyses - section 4.1.3
  - Effectiveness analyses - section 4.1.4
  - Assessment criteria - section 4.1.5
  - Ranking of candidate solutions - section 4.1.6
- The process approach - Chapter 4.2, which includes:
  - The location of the process in the development cycle - section 4.2.1
  - The purpose, inputs and outputs of the process - section 4.2.2
  - The description of the activities of the process - section 4.2.3
  - The ontology elements - section 4.2.4
  - The verification and validation of assessments - section 4.2.5
  - The artefacts generated by the process - section 4.2.6
- The practice - Chapter 4.3, which includes:
  - Modelling techniques and methods - section 4.3.1
  - Expert opinion and interviews - section 4.3.2
  - Evaluation example - section 4.3.3
  - Recommendations and FAQ - section 4.3.4

4.1 Assessment concepts and principles

Purpose of the system properties assessment

The system properties assessment activity evaluates, as objectively as possible, the engineering data in order to choose the solution, or the most relevant solution elements. It enables the development of a coherent and accurate set of engineering data, including system requirements, architectural characteristics and design properties.

During engineering, assessments are carried out whenever decisions or technical choices need to be made or justified, not only to compare different architectures, but also to justify agreement or disagreement with system requirements. The properties assessment activity provides a rigorous approach to technical decision-making. It performs trade-off studies based on costs analyses, technical risks analyses and effectiveness analyses.

> "Evaluate and choose" are two of the main tasks of the specialist in systems engineering.

4.1.1 Trade-off studies

One of the major concerns of development is to ensure that all the technical data of a system (essentially stakeholder needs, system requirements, and design properties) is consistent and that the values of that data are relevant. Trade-off studies are central to this issue.

The trade-off studies consist of analysing, evaluating and comparing the characteristics of each candidate solution (or candidate engineering data sets) to determine which of the candidate solution(s), as a whole, is closest to the imposed criteria. The various characteristics studied are grouped into three main types of criteria:

- cost
- risk
- effectiveness.

Each of these criteria is the subject of a specific study in four areas: **definition of the scope** of the study, **definition of comparison scales**, **quantification** of candidate solutions or engineering data sets, and their **ranking**.

In the context of an R & D project, trade-off studies are **planned**; that is to say, at what times these studies will be carried out, with what methods, resources and skills.

Definition of the scope of the study

The scope specifies and limits the properties, characteristics, criteria and/or sub-criteria, which will be taken into account in the study. To define the scope, one must look for significant or relevant parameters. For example:

- What costs should be taken into account?
- What are the potential and acceptable technical risks?
- How to characterize the effectiveness of the system? Performance, safety, security, etc.?

- What elements or engineering data should be studied: stakeholder requirements, system requirements, logical architectures, physical architectures, design elements, components, etc.?

Definition of comparison scales

Scales of value are needed to quantify properties, characteristics or criteria, and to make comparisons. Their definition consists of identifying the upper and lower limits, as well as the type of evolution (linear, logarithmic, etc.) of each property, characteristic or criterion.

For example, the reliability is linear from 0 to 10, with 10/10 corresponding to less than one failure per year, and 0/10 to more than one failure per week.

Care must be taken to define coherent scales: for example, a rating close to 0 will always reflect a negative aspect, and a rating close to 10, a positive one. Similarly, it is possible to choose from 0 to 1, from 1 to 10, from 0 to 100, etc., but all characteristic or properties must use the same interval.

Quantification of criteria and solutions

Quantification assigns a rating, or relative weight, to each property, characteristic, criterion or sub-criterion relative to the others.

The quantification of the criteria can be expressed:

- in **absolute** terms; for example, the maximum cost of the product is 100 k Euros, the minimum cost is 50 k Euros;
- in **relative** terms; for example, the cost will make 30% of the choice, efficiency 35%, risk 20%, aesthetics 10% and company policy 5%.

Each candidate solution, or engineering data, is positioned on the scale of comparison for each criterion.

The difficulty of the trade-off study is to quantify the three criteria of cost, efficiency and risk on identical bases to make valid comparisons. Indeed, if the candidate solution # 1 is quoted in Euros and the candidate solution # 2 in Dollars, the comparisons become complicated. Quantification is a complex operation and requires the use of models.

Ranking of solutions and optimisation

It is possible to establish a **score** for each candidate solution by combining its position on the comparison scale of each criterion with the relative weight of each criterion. The overall ranking of the solutions is obtained in relation to this score.

The optimization of system properties is obtained by improving the score of some interesting solutions, modifying either a requirement or an architectural characteristic, or by combining elements of different solutions to define a new solution.

Planning of trade-off studies

The trade-off study is one of the activities of the system properties assessment process. This process is executed iteratively throughout the project. The trade-off study is thus carried out several times, with an improvement in the accuracy of quantifications over time; one starts "coarse grain" when one has only the stakeholder requirements, then one refines as the engineering data become more precise throughout the project, using the system requirements, then the architecture elements, then design elements. One speaks of an optimal solution when the evaluations are more precise and they concern only the still remaining solutions - see Figure 18.

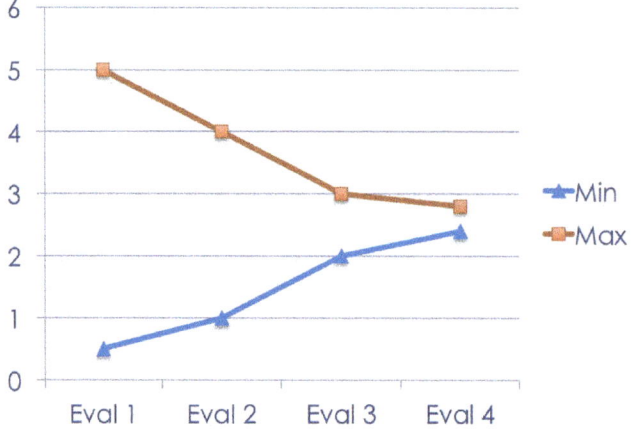

Figure 18 - Evolution of quantifications

Limits of studies

Trade-off studies collect relevant information to justify choices, and thus to make technical decisions. But, a decision-making process is not an exact science: trade-off studies have limits. Indeed, it is necessary to take into account:

- **subjective** variables; for example, the car must be beautiful ... but what is a beautiful car?
- **uncertain** data; for example, to calculate the maintenance costs of the manufacturing chain over the entire production period, inflation must be taken into account ... but what will inflation be in 5 years?
- **sensitivity** analyses; an overall assessment score associated with each candidate solution is not absolute; it is recommended to check the robustness of the decision model by conducting a sensitivity analysis, which involves stimulating this model by small variations in the values of the assessment criteria (e.g., the relative weight); the choice is robust if the variations do not modify the ranking by the scores.

A relevant trade-off study should specify the assumptions used, the variables or criteria taken into account, and the confidence intervals of the results presented.

Often the decision-maker has to make strategic choices, committing the bulk of the budget, long before they have certainty about the system engineering data; effectiveness in particular. Indeed, it is necessary to know that:

- 50 to 70% of a budget is locked by the choice of system architecture,
- about 90% of the budget is locked by the choice of the design of the system elements (components).

The existence of well-known experiences (Knowledge Management), which has enabled the construction of important databases and models, whose efficiency has been demonstrated, is a considerable advantage.

Chapter 4.1. System properties assessment: Concepts and principles

Method and synthesis

The method for conducting trade-off studies is summarized in Figure 19.

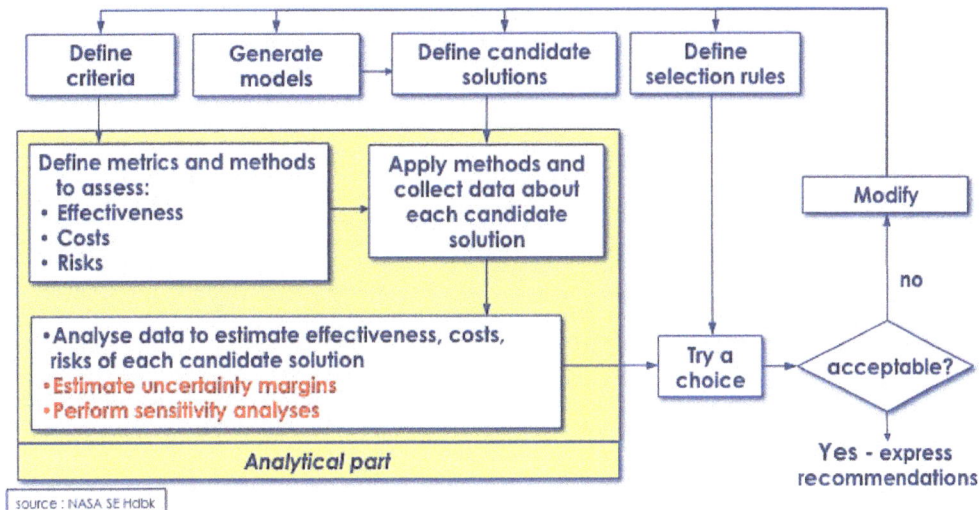

Figure 19 - Method for conducting trade-off studies

The method consists of:

- Defining assessment criteria
- Generating models of each candidate solution that allow for collection of data
- Defining rules of choice, such as algorithms, comparisons with weighted criteria, multi-criteria analysis models combined with specialist expertise
- Defining metrics and methods to evaluate cost, risk and effectiveness criteria and sub-criteria
- Applying these methods to collect metrics and data for each candidate solution
- Analysing these data to estimate the costs, risks and effectiveness of each candidate solution
- Estimating the margins of uncertainty
- Performing sensitivity analyses on the relative weights of the criteria
- Make a potential choice among the candidate solutions, checking the following aspects of this solution:
    - Are objectives, constraints, requirements satisfied?
    - Are subjective aspects included in the study?
    - Is the choice of the solution robust (sensitivity)?
    - Is a refinement of analytical analyses (metrics) necessary to differentiate the selected solution from the others?
- If the choice is acceptable, make recommendations and justifications based on the various analyses performed. Otherwise, review and modify the criteria, or models, or solutions, or rules of choice until a decision is possible.

4.1.2 Cost analysis

> **Scope of cost analysis**

Cost analyses are used to evaluate, estimate or calculate the cost of tangible items. When the engineering of a system is sufficiently advanced, as soon as it is possible to allocate functions on concrete system elements (components), these analyses may concern:

- Candidate physical architecture cost
- The cost of activities of a program life cycle phase
- The life cycle cost (or total ownership cost)

The cost of a physical architecture of a system can be easily estimated from the cost of each component (via component catalogues and/or cost bases), as well as the physical interfaces or links that connect the components of the system, or those connecting the system to the external elements. The calculation is a simple sum insofar as the elements that make up the cost are fixed; which is only true at a given moment in the life of the system.

The cost of activities may relate to a part of the life cycle, such as development, production, transfer for use, operation, maintenance or logistical support, disposal. The following lists of elements to be taken into account for the estimate are not exhaustive.

- **Studies and development cost**:
  - workforce, modelling and simulation tools, mock-ups and prototypes, test means, tests, documentation, training of developers, procurement, project management, etc.
- **Production cost**:
  - infrastructure investments, machinery, tools and premises, workforce, training of operators, raw materials and transport of these materials, spare parts and storage, resources (water, energy, etc.), risks and nuisances, disposal, treatment and storage of waste, structure costs (taxes, management, purchases, documentation, quality controls, cleaning, regulatory controls, etc.).
- **Transfer for use, sale, after sale cost**:
  - packaging and storage, transport and delivery, fitting out of premises or installations, adaptation of the environment, documentation to be provided, spare parts stocks, warranty costs, structure costs (shops, workshops, etc.), training, workforce, etc.
- **Operation cost**:
  - workforce, training, resources needed to operate the product (consumables, water, fuel, energy, etc.), disposal, treatment and storage of waste, risks and nuisances, structure costs (taxes, management, purchases, documentation, quality, cleaning, etc.)
- **Maintenance or logistic support**:
  - workforce, training, preventive actions, maintenance, repairs, spare parts, storage, regulatory controls, etc.

- **Dismantling or disposal cost**:
    - workforce, training, disassembly, disposal, treatment and recycling of components and waste, resale, final storage, etc.

Life cycle cost (or total ownership cost) refers to the costs over the entire life of the system (product, service, organization). It is a dynamic view of costs, because we are interested in the variations of the elements that make up the cost of the system throughout its life, according to the evolutions of the environmental, societal, economic conditions, etc. It allows an overall view on parameters that are often related, or to identify interactions; for example, lowering development costs, maintenance costs may increase.

It is recommended to present cost models that take into account the distribution of costs over time, as illustrated by the curve of Figure 20. In the estimate and partitioning of the life cycle cost, do not forget to include the gains from mid-life resale, inflation or deflation.

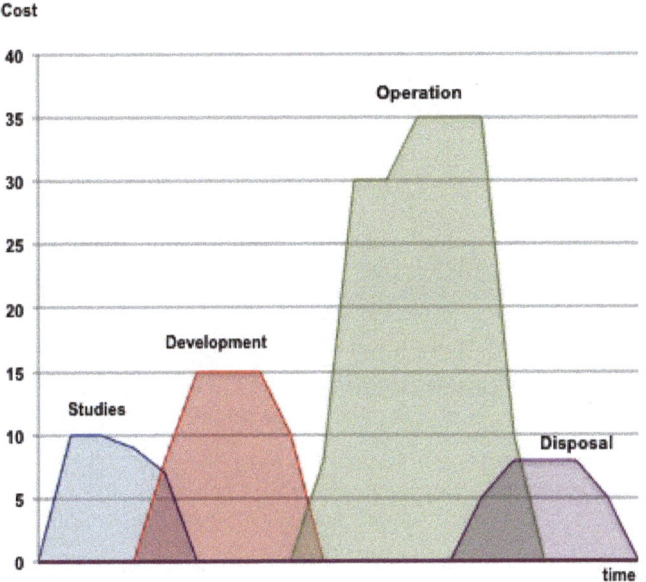

Figure 20 - Example of life cycle cost partition in time

Comparison scales for costs

Comparison scales for costs are often established in a linear manner, as shown in the diagram of Figure 21.

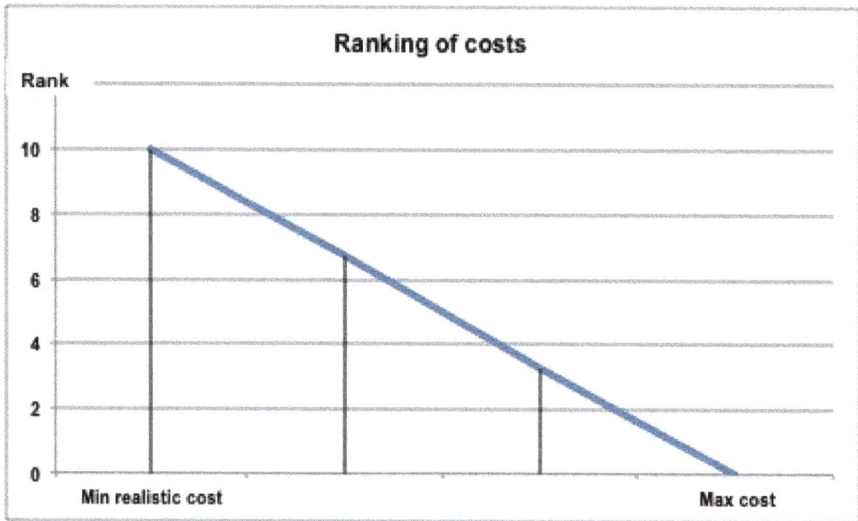

Figure 21 - Example of a conversion cost / estimate scale

Quantification of costs

Cost hypotheses

Cost is a seemingly objective criterion. Nevertheless, a cost is not just an **amount** in a monetary unit. The evaluation of a cost depends on a number of parameters:

- From the **composition** of the system into components (products and services); therefore, costs of each component that may vary over time. Each **amount**, for each constituent, is associated with a confidence interval, a reference date or period, the source or evaluation technique.

- From involved life cycle activities; i.e. to **develop** the system, **produce** the components and integrate them, **transfer** the system into operation, **operate** it, **maintain** it in operational conditions and even **dismantle** it.

- From variation over time due to changes in the environment (economic, social, contractual, geopolitical, etc.) that may impact the entire system. The cost of each phase of the life cycle is therefore accompanied by the **confidence interval** (minimum, maximum, average), the **reference period** (date of the estimate), the **assessment technique** or the sources consulted.

- From financing on investments, borrowings, exchange rates (financial cost), or investment income (financial income).

- From risks that may arise throughout the life of the system (e.g., defective or delayed supplies, accidents, etc.) and solutions to reduce these risks (e.g., insurance).

If relevant cost elements are desired, the cells in Table 1 must be filled in.

Cost	Hypotheses			
	Amount	Interval of confidence (min, max, probable)	Date / reference period	Source (assessment technique)
Development				
Production				
Operation				
Maintenance / support				
Dismantling				
Finance				
Risk				

Table 1 - Life cycle cost items

During the first quantifications, it is not possible to carry out a precise quantification, since the studies of architecture and design are not sufficiently advanced. However, the first arbitrations will commit the major part of the budget; it is therefore important to establish realistic budgetary envelopes as early as preliminary studies. It is necessary to identify and assess or estimate the factors that are important sources of expenditure; for example:

- innovative or little-known technologies
- the use of specific processes requiring a high level of accuracy
- important special tests
- high level of dependability (availability, safety, survivability)
- constraining environments
- regulatory or legislative constraints

These global quantifications are based on previous studies. It is possible to use "learning curve" or "beta curve" (see section 4.3.1.2), based on previous statistics.

Detailed quantifications are reserved for the latest evaluations, when the engineering is sufficiently mature. They serve to refine the differences between the remaining candidate solutions. These studies quantify each cost item using and summing-up the data from the assessments of the subsystems and components that make up the system. These studies are long and tedious and require advanced maturity of engineering.

Cost studies should be updated as the project progresses and compared to the target budget. In each study, the amounts already spent are recorded but treated as "completed"; the choices to be made relate only to the amounts to come, without taking into account what has been spent previously.

Chapter 4.1. System properties assessment: Concepts and principles

Value analysis

Value analysis is a method of analysing an existing product or system in order to define a new product or system on the basis of the existing one, but performing better technically or economically or socially, depending on the acceptance the term "value" is given. It can therefore be applied to the assessment of candidate solutions (physical architectures) during the initial definition of the system. The idea is to "quantify" the value of the functionalities of the system with respect to a given allocation of functions on physical components (i.e., an architectural solution). It is then possible to compare different allocations (i.e., different architectural solutions).

For an architectural solution, the method is the following:

- ♦ Identify the components of the architecture that participate in a functionality (or service) of the system.
- ♦ Estimate the participation rate of components in the functionalities.
- ♦ Estimate or calculate the price or cost of each component.
- ♦ Distribute the cost of each component to the functionalities by multiplying the cost of the component by its participation rate for each functionality; the sum gives a "cost" to each functionality for this architectural solution.

Table 2 illustrates a hypothetical example of calculation of the cost of the functionalities (services) of the system, for a given architecture composed of six components that perform four services. The table is read as follows:

- Lines: component C1 participates to function F1 for 50% (50) and to F3 for 50% (50). Same thing for each line.
- Columns: function F1 use 50% of C1 (50) + 40% of C3 (60) + 50% of C4 (75) + 10% of C5 (30) = 215. Same thing for each column.

		Functionalities				Cost of components
		F1	F2	F3	F4	
Components	C1	50%		50%		100
	C2		20%	10%	70%	200
	C3	40%		60%		150
	C4	50%	10%	40%		150
	C5	10%	80%		10%	300
	C6		50%		50%	100
Cost of functions		215	345	220	220	1 000

Table 2 - Estimate of the "cost" of the functionalities of the system

Reduction of costs

Reducing the cost of a system can be achieved by acting on the various components of the cost: on system elements (components), on life cycle activities, on financing of resources, and so on.

Value analysis makes it possible to consider cheaper components choices, or to reconsider the physical architecture by partitioning the functions in different ways and by allocating these partitions to different components.

It is also possible to reconsider the business model resulting from the first market studies (see the business or mission analysis process - Chapter 4.2, Volume 2 of this series, Systems Opportunities and Requirements [Faisandier2 2013]), i.e. how to sell the system. For example, instead of selling a product at a price that is not readily accepted by potential users, the company may retain ownership of the product and sell a service that includes the availability of the product, support for use, and maintenance; which, on the other hand, involves setting up financial engineering within the company.

Nevertheless, it must be understood that the reduction of costs may lead to less efficiency and more risks. Therefore, the cost reduction study must be included in a more comprehensive exercise to optimize all operational, risk and cost parameters - see section 4.1.6.

4.1.3 Risk analysis

Risk paradigm

Figure 22 presents the items engaged within risks analyses.

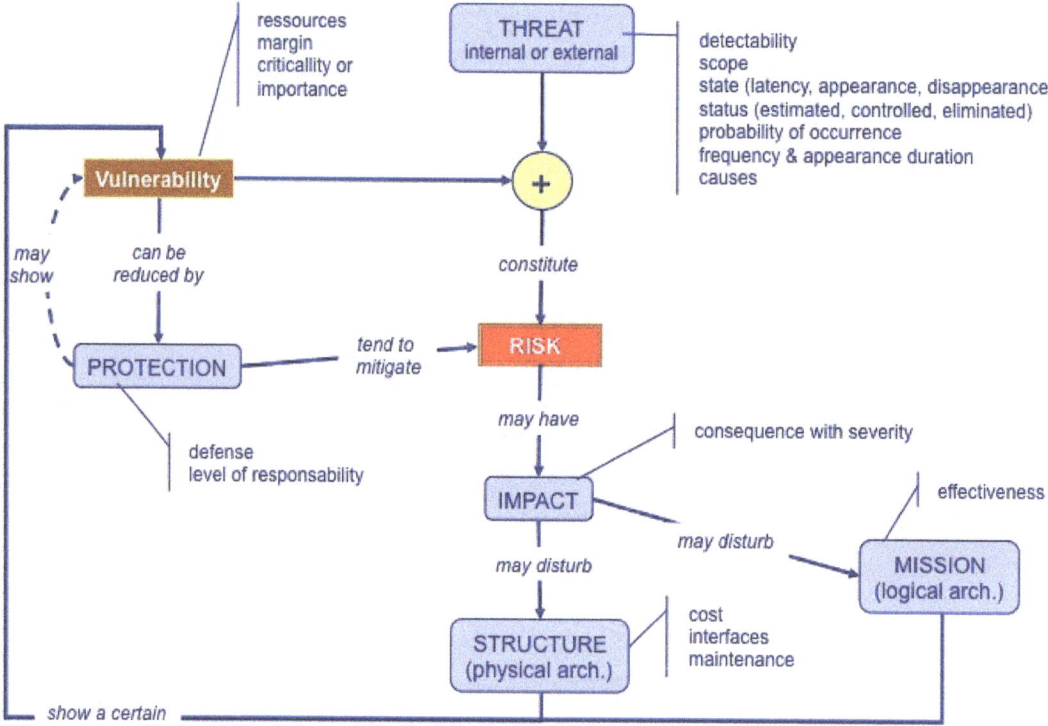

Figure 22 - Paradigm of the risk

The reading of this figure is as follows:

- A **risk** is the combination of a **threat** and a **vulnerability** to this threat.

- A **threat** (or hazard) arises from events, or a combination of events that are more or less inappropriate, internal or external to the system. A threat is characterized by its *detectability*, *scope*, *state* (latency, appearance, disappearance), *status* (estimated, controlled, eliminated), *probability of occurrence*, *frequency* and *duration of appearance*, and *causes*.

- There is a risk only if there is a **vulnerability** to a threat or hazard. According to this vulnerability, the threat creates a **risk** that can have more or less serious **impacts** on the **mission** of the system (logical architecture - effectiveness) and/or on its **structure** (physical architecture - costs, failures, etc.) and consequently on the use of the system.

- The components of a given architecture always present a certain vulnerability. This vulnerability can be reduced by **protections**, which tend to mitigate the risk; a protection is characterized by a *defence*, a *level of responsibility*.

Consequently, in engineering, any study of technical risks is based on the following actions:

- Analysis of **potential threats or hazards** (natural disasters, maliciousness, breakage of materials, etc.), their probability of occurrence, and their detectability.
- Analysis of the **vulnerability** of candidate solutions (architectures and design) in relation to these threats and hazards.
- Quantification or assessment of the corresponding **risks**, in terms of criticality and acceptability according to the **impacts** (or consequences) if the risk is proven.
- The introduction of **protections** or **preventions** (risk reduction actions) to mitigate the importance of risks to acceptable values, or to eliminate them.

Scope of risk analysis

This section presents examples of risks; a list of risks must be established according to the nature of the system, its context of use (operational and environmental conditions) and the stage of the life cycle concerned (development, operation, maintenance, etc.). Three categories of risks are considered below: technical risks, external risks and management risks.

Technical risks arise when the system can no longer satisfy its technical requirements. They are usually expressed in terms of inadequate effectiveness or operational efficiency. These risks, or their causes, can be, for example:

- failure of components, obsolescence of components to be replaced, non-mastered technologies or innovations, non-compliant components,
- design errors, operator errors, rejections by users, handling errors,
- delays in supply, inoperative settings, inadequate procedures, malicious acts.

External risks relate to changes in the context or environmental conditions over which the project or company does not operate; for example:

- natural events: thunderstorms, floods, frost, heat, etc.
- political decisions or events, legal changes, court decisions,
- changes in policy direction.

Management risks may deal with:

- budgets: inadequate evaluation of the necessary means, delays, contractual clauses, breach of contract, failure of a customer or supplier,
- planning: poor evaluation of task duration, forgotten tasks, unexpected obstacles, etc.

Notes :

Technical risks must not be confused with project risks; although the method of analysing and managing them is the same. Technical risks address the system itself, not its development. Obviously technical risks can react on the risks of the project.

The risks may be interrelated: for example, a change in legislation may require the resumption of a study, which may entail additional costs and increased delays; or lightning may lead to the destruction of electronic control boards and the shutdown of a production.

Chapter 4.1. System properties assessment: Concepts and principles

Comparison scales for risks

Risks can be assessed in different ways; in relation to the severity of the impact (consequence) if the risk is proven, or relative to the acceptability of the risk, which takes into account the probability of occurrence of the risk and the severity of the consequence.

Severity scales are used to classify risks according to their potential consequences. For example, the Beaufort scale, which classifies the winds from 0 (calm) to 12 (hurricane) on the basis of the environmental consequences; the scale of nuclear accidents, which classifies incidents from 1 (operating anomaly) to 6 (major accident) depending on the releases and the consequences on the population.

Severity scales are generally logarithmic. For example, it is possible to define a logarithmic risk scale for planning or delay (Figure 23): 10 for no delay, 9 for 1 day, 8 for 2 days, 7 for 4 days, 6 for 1 week, 5 for 2 weeks, 4 for 1 month, 3 for 2 months, 2 for 4 months, 1 for 8 months and 0 beyond.

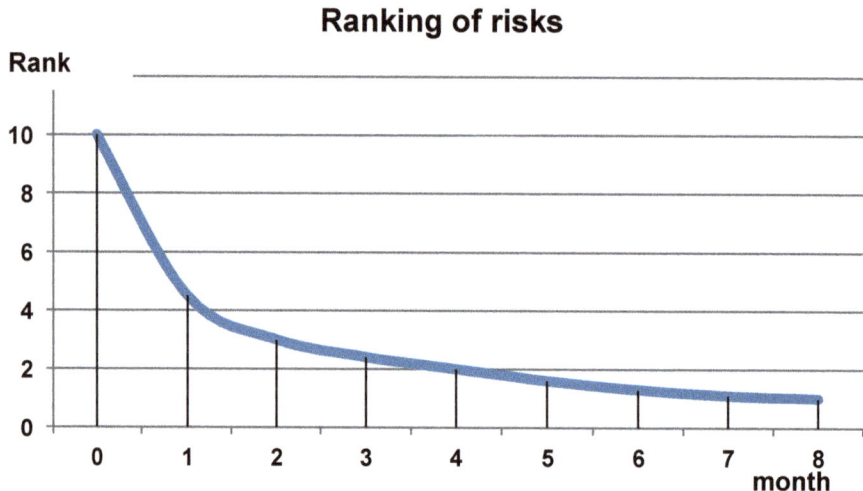

Figure 23 - Example of a severity logarithmic scale for a planning risk

Acceptability scales take into account both the severity of the consequences and the probability of occurrence of the risk. To take a risk means that one accepts the risk comes and that one takes the responsibility and assumes the consequences. Indeed, one can accept the repetition of minor incidents, or a more serious incident that has a low probability of occurring, but one will not tolerate a major risk with a high probability of occurrence. Figure 24 illustrates an acceptability diagram.

Chapter 4.1. System properties assessment: Concepts and principles

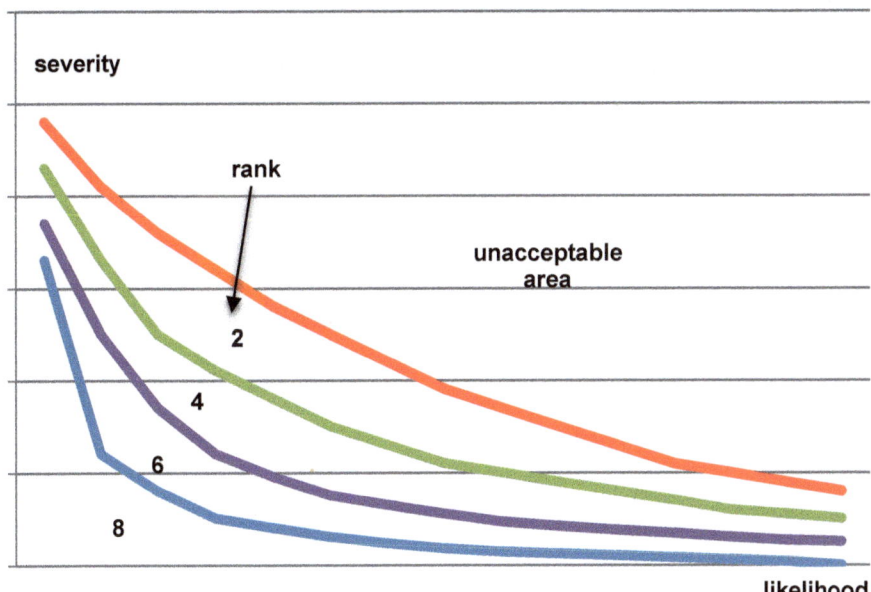

Figure 24 - Example of a risk acceptability diagram

Acceptability can also be represented in tables that cross the probability of occurrence and the severity of the consequences; each cell in the table contains a description of the action envisaged to accept, mitigate or eliminate the risk. Figure 25 illustrates a table with nine cases of criticality. Each industrial sector uses its own definitions of risks and criticality; the tables usually have 16 or 25 cases of criticality.

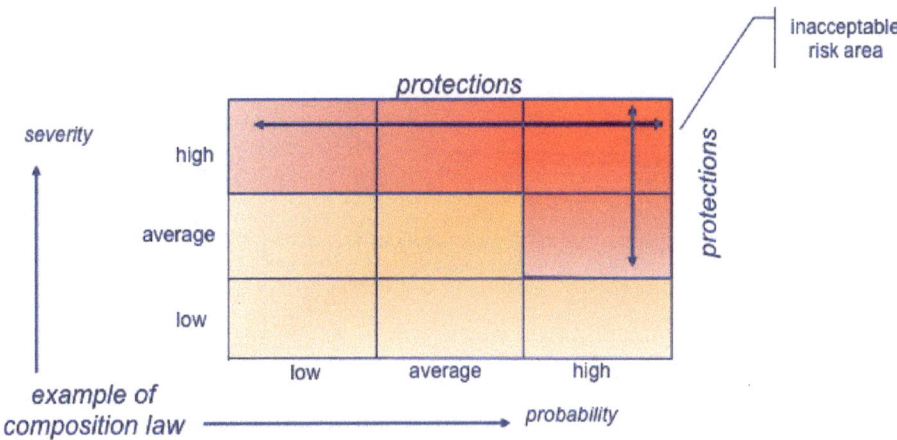

Figure 25 - Table with nine cases of risks criticality

Chapter 4.1. System properties assessment: Concepts and principles

Quantification of risks

A risk is quantified by a level or **criticality**. Risk criticality is a measure or value resulting from a composition of the risk characteristics, mainly its probability of occurrence and the severity of the impact, or consequence on the system itself, on the assets and persons of the environment.

Because of the indeterminism of hazards and threats, there is no exact mathematical law to express the criticality of risks; it is for this reason that the probabilities are used.

On the other hand, probabilities require the use of existing data, in sufficient numbers to deduce a probability law. Most manufacturers and operators have fault statistics on their equipment or products. Natural hazards are the subject of public studies; for example, fire specialists may calculate a probability of risk in a room, etc.

In the absence of precise data, an evaluation can be made from neighbouring elements whose probability of occurrence is known, and by increasing the risk as a caution.

Mitigation of risks

In summary, a risk:

- is predictable or not (detectability),
- has a probability of occurrence,
- may have impacts, or consequences, on the system, and/or on external assets and people, characterized by severity levels,
- is acceptable or not depending on its criticality, which is composed of:
    - constraints or elements of the context of use,
    - the probability of occurrence of the risk,
    - the severity of consequences.

From the level of acceptability of the risk (criticality), it is decided to put in place protections or specific actions to mitigate the risk. These protections may be of different types:

- Technical protections - for example, the installation of a redundant line, containment of errors or failures, reinforcement of material thickness, preventive maintenance actions;
- Human protections - for example, adequacy of staff skills, training;
- Procedural protections - for example, qualifications, check points in a process, prohibition of passage in a lane.

4.1.4 Effectiveness analyses

Effectiveness analyses are used to characterize, during its engineering, the operational efficiency of a system using its various models, in particular candidate architectural models. These analyses mainly concern **performance** and **operational conditions** or constraints.

For a given system, it is therefore:

- to identify a set of property coming from non-functional requirements, architectural characteristics and design properties, in order to compare candidate solutions,
- to develop the most appropriate models for the properties that characterize the operational efficiency,
- to define assessment methods.

Scope of effectiveness analysis

The list of effectiveness studies is based on the categories of requirements - see Chapter 6.2.6 Volume 2, Systems Opportunities and Requirements [Faisandier2 2013], but more importantly on architectural characteristics and design properties as defined in Annexes 3, 4, 5 of Volume 3, Systems Architecture and Design [Faisandier3 2013].

The following list is intended to be exhaustive and seeks to shed light on the candidate solutions from different views. But it is essential to limit the study to the most important properties for the system under consideration. The choice of the properties to be assessed is one of the main difficulties of the effectiveness analyses. For example, if the product is a one-shot-product, its maintainability and adaptability to other missions are not significant criteria.

Analyses related to operation

Performance analysis is essential. We analyse:

- the ability of the system to satisfy its mission(s), the performance efficiency of each function of the system
- the reaction durations of the system to the various stimuli or commands
- the behaviour of the system when it is submitted to unforeseen disturbances or limitations
- the adaptability of the system, in the case of learning systems or adaptive systems

Behavioural analysis deals with the system in its operational context of use:

- connections or exchanges with elements or other systems of the context (interoperability)
- operational modes and transition between these modes
- distribution of weights in static (mass balance), or in dynamic
- consumed, produced or recovered power or energy

Human factors and ergonomics studies:

- relations or interfaces between the system and users, operators, maintainers, carriers, inspectors, etc.
- complexity of tasks, level of stress
- working conditions (location, noise, heat, light, etc.)

- the level of knowledge that must be reached by the various players in the system
- training program, etc.

Dependability analyses

These analyses help to identify critical components of the system that may be faulty, the causes and consequences of these failures. We can study:

- **availability** of the system that includes the **reliability** of the components regarding failures (failure rate over a given period of time) and **maintainability** (characterized by downtime for repair, correction, maintenance, administrative procedure, supply of components, etc.); these studies are generally performed with the help of existing statistics
- **safety** and **security** of the system regarding its vulnerability to threats, aggressions, attacks, or hazards
- **survivability**, that is to say the way the system behaves in a hostile environment, and its ability to continue the mission in a degraded mode, or abandon it or reconfigure itself; these analyses are essential in certain fields such as nuclear, chemical, defence; but can be studied in any type of system

Environmental conditions analyses

These analyses deal with:

- **influences on the system** such as climatic conditions, electromagnetic compatibility (radio spectrum, etc.)
- **the impact of the system on the natural environment**, such as nuisances, waste disposal and means of disposal, etc.

Production analyses

These analyses deal with:

- **industrialization** or manufacturing process: ease of production, cycle time, manufacturing ranges, etc.
- **assembly**: mounting order, mounting methods, quality checks, etc.

Support or maintenance analyses

These analyses deal with:

- **check points**, tests and corresponding means to be implemented throughout the life of the system to guarantee its quality and conformance (performance / efficiency)
- technical means to **maintain** the system in its working condition at the best cost: maintenance plan, stock of spare parts, standardization, ease of supply, disassembly and reassembly of components, handling equipment, diagnosis tools, etc.
- **transfer for use** or start-up of the system: preparation of the installation site, connections to the various networks (communication, energy, water, etc.), packaging, transport, handling, storage, stocks of initial spare parts, initial training, regulatory controls, etc.
- **disposal** or dismantling of the system: disassembly, recycling of materials, re-assigning of personnel, final storage, cleaning of the site, etc.

Comparison scales for effectiveness

These scales have no particular difficulty: for each characteristic or significant property, one can define a minimum value and a maximum value, and mark linearly from 0 to 10 (or 1, or 100) between the two extremes.

Quantification of effectiveness

Each characteristic or property has its own units, its own representation models. Simulations (see section 4.3.1.1), calculations, use of models, expert opinion (see section 4.3.2) can be carried out.

Quantifying globally the effectiveness of a system is a rather complex task, because each candidate solution is a combination of characteristics. Only a multi-criteria approach can account for the systemic aspect of effectiveness (see section 4.1.5).

Effectiveness improvement

Each characteristic or property of the system can be the subject of an optimization study carried out by the concerned specialists.

But, keep in mind that the optimization of the effectiveness of the system is not necessarily obtained by the optimization of each of its properties; the properties react positively or negatively on each other - see Figure 70 on the mutual influences between life cycle aptitudes, Volume 3, Systems Architecture and Design [Faisandier3 2013].

4.1.5 Assessment criteria

Decision model

A decision support model (or decision model) must take into account the multiple selection criteria resulting from the previous analyses, as well as their combination. The projection of the valued results on a single axis facilitates the decision-making (for example by calculating a score).

The establishment of criteria must precede any calculation or simulation; in fact, knowledge of the results could influence the choice of criteria, which would distort the approach. To avoid the phenomenon of coupling between the criteria, it is necessary to take criteria as independent as possible from each other.

The three key variables of **cost**, **risk** and **effectiveness** generally limit the scope of the solution by imposing:

- maximum cost,
- minimum efficiency,
- maximum risk.

The solution domain makes it possible to exclude the candidate solutions that are not part of it (too costly, too time consuming, too risky) - see Figure 26.

For most complex systems, these three variables are related; for example, if the cost is reduced, the risk increases at constant efficiency.

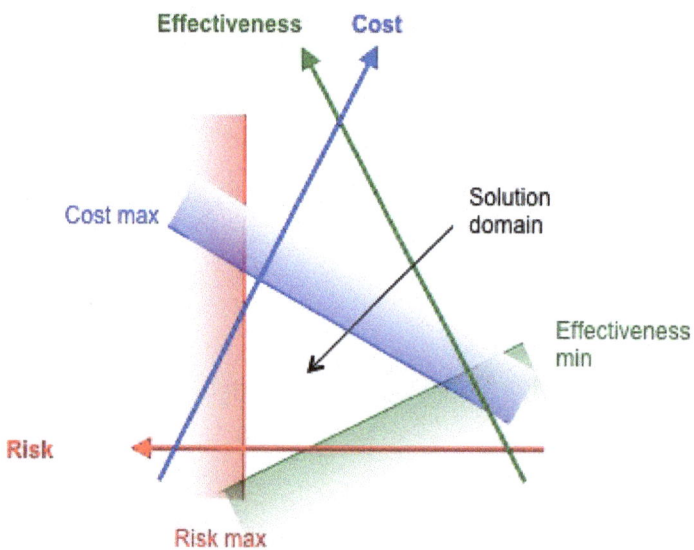

Figure 26 - Delimitation of the solution domain

In a decision support model, the number of assessment criteria may be limited to a few characteristics. For example, the solution with the least cost for a given level of risk, or the one that is most effective for a defined budget, can be selected.

However, it may be necessary to take into account more criteria, in particular in a search for an optimized solution over a set of requirements or architectural characteristics or design properties. In this case, it is better to organize the assessment criteria among themselves, prioritizing them in the form of a tree of criteria and sub-criteria, and then assigning a relative weight to each criterion in relation to the others. Relative weighting is based on expert opinion or stakeholder interviews - see section 4.3.2. Below ,Figure 27 illustrates such a weighed tree.

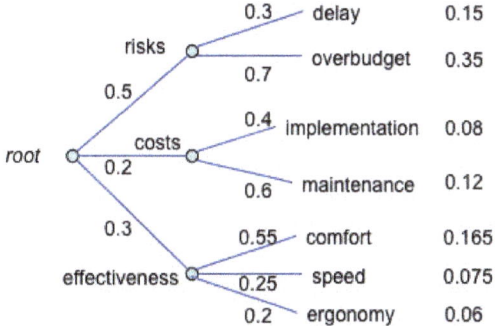

Figure 27 - Weighed criteria tree

In this figure, the criteria for risk, cost and effectiveness are composed of sub-criteria. The sum of the **relative weights** of each node of the tree is equal to 1. An **aggregate weight** is defined for each leaf criterion by multiplying the relative weights of each of the criteria on the branches to the first node of the tree. The sum of the aggregate weights is equal to 1. The aggregate weights represent the systemic importance on all the criteria and sub-criteria. It is then sufficient to compare the candidate solutions only with respect to the leaf criteria of the tree.

Another example of a multi-criteria model for evaluating candidate system architectures is shown in Figure 28.

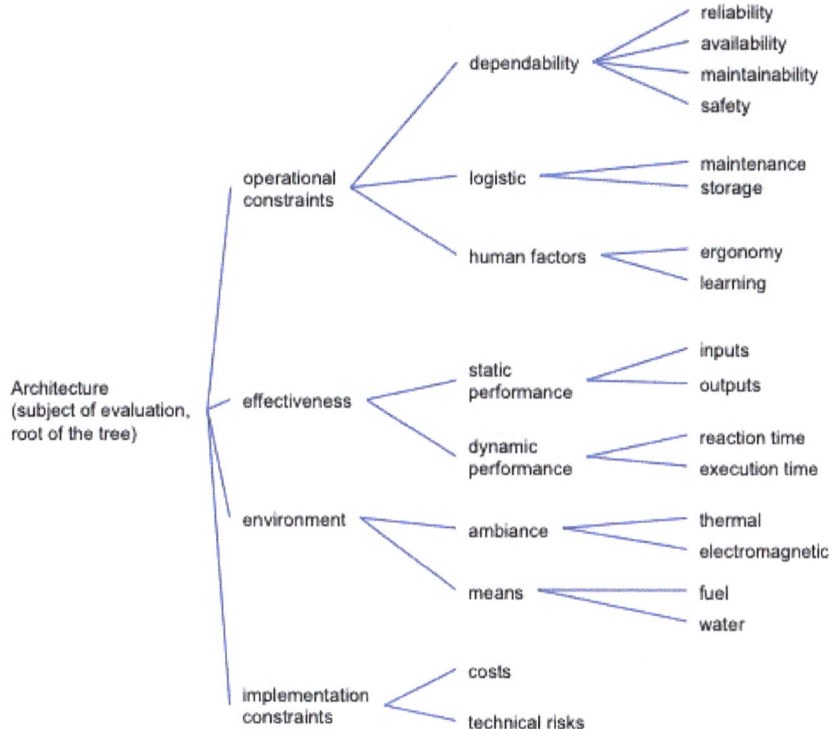

Figure 28 - Example of a multi-criteria model

Chapter 4.1. System properties assessment: Concepts and principles

4.1.6 Classification of candidate solutions and optimization

Each candidate solution is analysed and assessed using the different methods described above regarding the cost, effectiveness and risk criteria; more precisely, the leaves criteria when the criteria are numerous and prioritized according to a tree structure.

The results of analyses and evaluations are presented in tables or graphs for comparisons. The assumptions taken into account must also be expressed.

The ranking of the candidate solutions can take into account, or not, the uncertainty of the results of the calculations or simulations.

Classification methods without taking uncertainty into account

In this case, the result of an analysis must be quantifiable and comparable from one candidate solution to the other. The simplest method is to multiply the result of the analysis or calculation with respect to each leaf criterion (mark) by the weighting of this criterion, and then add the marks to obtain a score.

Taking the criteria in Figure 27, the evaluation of each candidate solution (mark of 0 to 1 given by calculation results, simulations or experts opinion) with respect to each criterion is presented below, in Table 3:

	Candidate solutions (mark)		
	S1	S2	S3
Delay risk	0.2	0.7	0.5
Budget risk	0.1	0.3	0.7
Development cost	0.6	0.2	0.7
Maintenance cost	0.2	0.5	0.8
Comfort	0.4	0.8	0.6
Speed	0.2	0.5	0.8
Ergonomic	0.4	0.6	0.5

Table 3 - Table of marks of candidate solutions

The comparative results are obtained by multiplying the **mark** by the **aggregate weight** of each leaf criterion (see Figure 27), and by adding the digits of each column, to give a score - see Table 4.

		Candidate solutions (mark)		
		S1	S2	S3
Leaves criteria	Delay risk	0.03	0.105	0.075
	Budget risk	0.035	0.09	0.245
	Development cost	0.048	0.016	0.056
	Maintenance cost	0.024	0.06	0.096
	Comfort	0.066	0.132	0.099
	Speed	0.015	0.0375	0.06
	Ergonomic	0.024	0.036	0.03
Score (mark x weight)		**0.242**	**0.4765**	**0.661**

Table 4 - Comparative table of the score of candidate solutions

For the selected criteria and their relative weight, the best solution is the candidate solution S3 (highest score).

Classification methods with taking uncertainty into account

To take uncertainty into account, the same method can be used as above, but using probability curves instead of the values of relative weights and/or marks. These calculations require the use of computer tools.

Another method consists in making a "minimax" choice: the lowest partial scores (weight x mark per criterion) of each candidate solution are searched (minimum), and the candidate solution with the least of lowest partial scores is chosen (i.e. minimizing the maximum score loss, hence the minimax name).

Robustness of the classification

Before making a final decision, it is necessary to verify that the classification obtained is robust by conducting **sensitivity analyses**. For example, if the scores obtained are close to each other, uncertainty or a slight change in the value of a relative weight or mark may cause the choice to be switched to another candidate solution; the choice is *weak*. To avoid this situation, the following arrangements must be made:

- define candidate solutions not coupled by values of characteristics that are too close
- choose the most independent criteria between them
- collect enough data to be able to correlate successive evaluations

When the choice is weak, it must be made more robust by changing the following points:

- ♦ add additional assessment criteria
- ♦ modify the candidate solutions to differentiate them more
- ♦ collect additional data or consider other models, etc.

Optimization of the selected solution

The optimal solution is to choose the solution that obtained the best score on all the assessment criteria, and then modify it, if necessary, by introducing the most positive elements resulting from the unsuccessful candidate solutions. This is possible, because each candidate solution was evaluated against several criteria and all the strengths and weaknesses of each candidate solution were highlighted.

In other words, optimization consists of increasing the score of the best solution by finely analysing its composition:

- Compared to system requirements:
    - Are all functions useful?
    - Is the efficiency too high or not high enough to meet the need?
    - Can efficiency be improved?
    - Are the constraints and operational conditions adequately met?
    - Are the margins on the different properties too large or insufficient?
    - Etc.
- Compared to architectural characteristics or design properties:
    - Accuracy, simplicity, availability, coupling rate, generality rate, independence, reaction duration, consumption, weights, dimensions, etc.
- Using experience feedback: can a technology, an assembly principle, an idea from an unsuccessful solution or another system not improve effectiveness, reduce costs, or mitigate risks?

Nevertheless, in the optimization actions, carefulness is recommended, because the best is sometimes the enemy of the good, and because engineering data are linked, a change can result in cost increases or efficiency reductions, and so on.

4.2 Process approach

This chapter describes:

- The location of the process in the development cycle - section 4.2.1
- The purpose of the process, its inputs and outputs - section 4.2.2
- The activities that compose the process - section 4.2.3
- The main ontology elements used - section 4.2.4
- The verification and validation actions to perform on assessments - section 4.2.5
- The main artefacts (documentation) to be produced - section 4.2.6

4.2.1 Location of the process in the development cycle

Figure 29 shows the location of the assessment process amongst the definition, implementation and integration processes.

The system properties assessment process is used iteratively when defining operational and technological concepts, and when defining the system, as many times as necessary to refine the definition of engineering data.

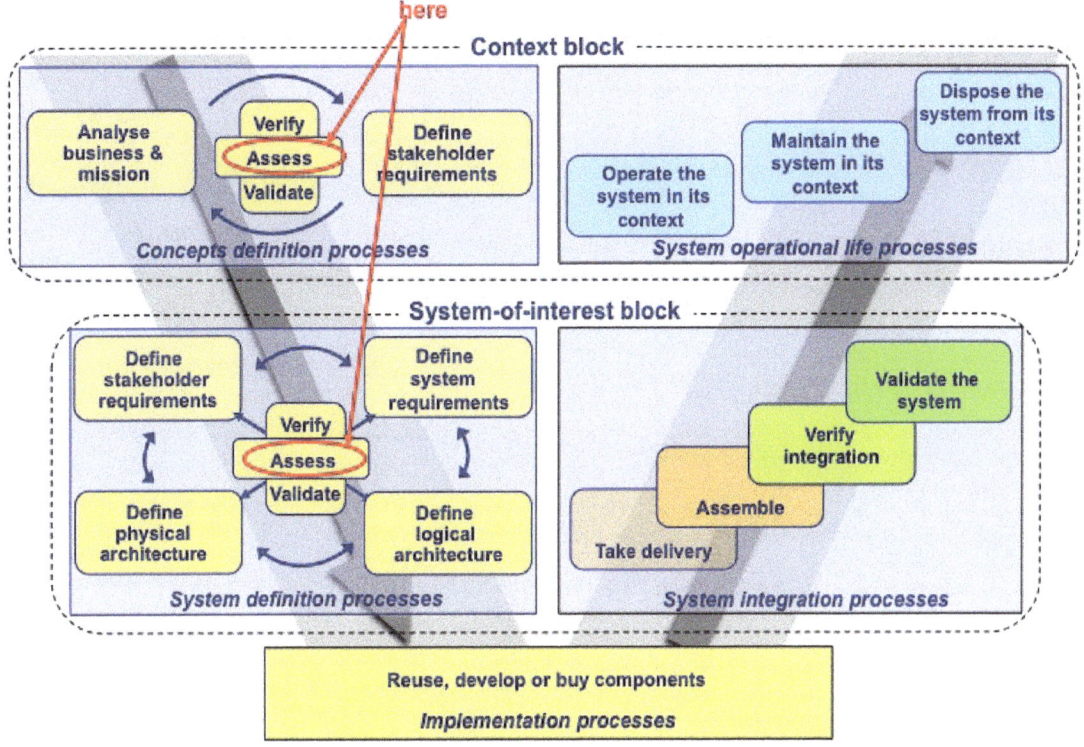

Figure 29 - Location of the assessment process in the development

4.2.2 Purpose of the process, inputs and outputs

Synthetic approach

The system properties assessment process (also called system analysis process) provides a rigorous approach to make technical decisions, solve conflicts between and select requirements, evaluate logical and physical solution alternatives, determine the progress of system requirements satisfaction, determine the pertinence of the set need-solution, support technical risk management. It allows one to make decisions only after evaluating costs, deadlines, effectiveness, and impacts on risks in the engineering or re-engineering of a system.

Inputs

The generic inputs of the process are the **candidate solutions**, or the engineering elements submitted to assessment.

Depending on the progress of the project, they are represented by a set of stakeholder requirements, or a set of system requirements, or logical or physical architectures through architectural characteristics or design properties.

As the process assessment is a support process, it can be used to assess other system development elements, for example, sets of verification and validation actions, sets of integration aggregates when verification, validation and/or integration strategies are established.

Outputs

The generic outputs of the process include:

- The selected solution elements (stakeholder requirements, system requirements, logical architecture, physical architecture, design elements)
- The results of the analyses and evaluations, the models of assessment criteria (decision models), the quantification models of the assessed properties, which enrich the justification document
- Removed solutions with their description and reasons for rejection to be included in the justification document
- The recommendations in order that technical decisions can be made
- Justified verification, validation, and/or integration strategies when the process is used in these cases

Chapter 4.2. System properties assessment: Process approach

Discussions

Support aspect of the assessment process

The system properties assessment process provides the necessary methods, techniques and tools for analyses: simulations, calculations, algorithms, various models, etc. All analysis, evaluation, justification and optimization tasks are grouped together in a process, because the engineering elements react with each other due to their relationships - see the generic engineering meta-model for system definition of the system in section 3.4.2.

The results of these actions lead to feedback on activities carried out previously, thus serve to consolidate the set of engineering data - see Figure 30.

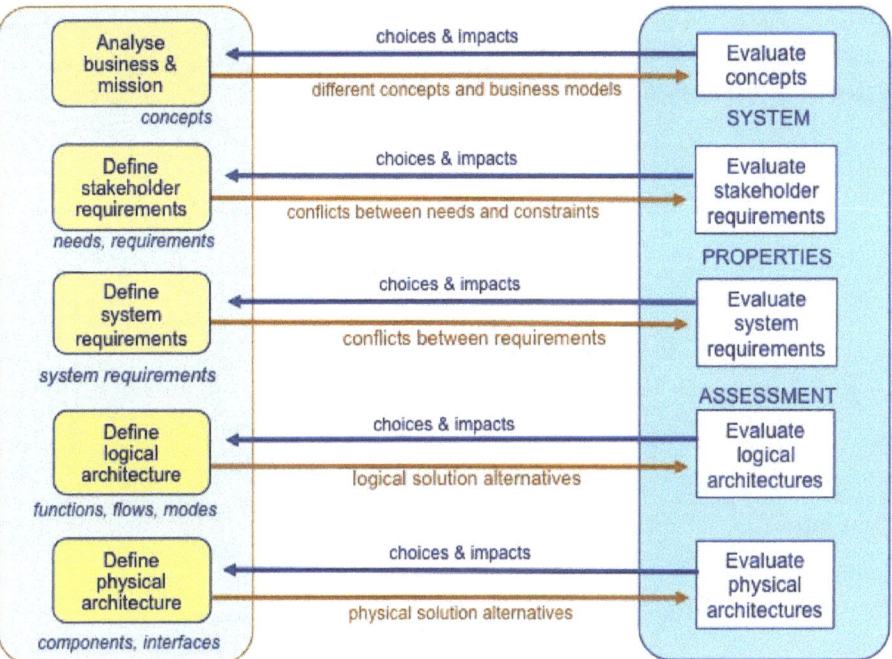

Figure 30 - Support aspect of the assessment process

Evaluation versus decision

The system properties assessment process gives an important place to the definition of the assessment criteria, in order to make technical decisions regarding the choice of the solution. Decision-making is not made by this process, but by the calling process. The assessment process provides the calling process with the results of various analyses and comparisons of candidate solutions against assessment criteria.

From an organizational viewpoint, it is important that the engineers in charge of these studies make recommendations for decision-making. Excluding them from the decision-making process would tend to disempower them, and consequently diminish their interest in the tedious tasks of analysis, evaluation, justification and optimization. This does not affect the use, or not, of a generic decision management process as defined in the standard ISO/IEC/IEEE 15288:2015 (Decision Management Process) [ISO 15288].

Chapter 4.2. System properties assessment: Process approach

4.2.3 Activities of the process

The main activities of the assessment process are the following:

   A. Establish a decision model composed of criteria
   B. Define engineering entities submitted to assessment
   C. Analyse elements submitted to assessment
   D. Perform trade-offs studies
   E. Rank the analysed elements
   F. Optimize the best elements
   G. Record results of analyses and justifications

Figure 31 presents the activities of the process, the engineering data exchanged, and main artefacts produced by the process.

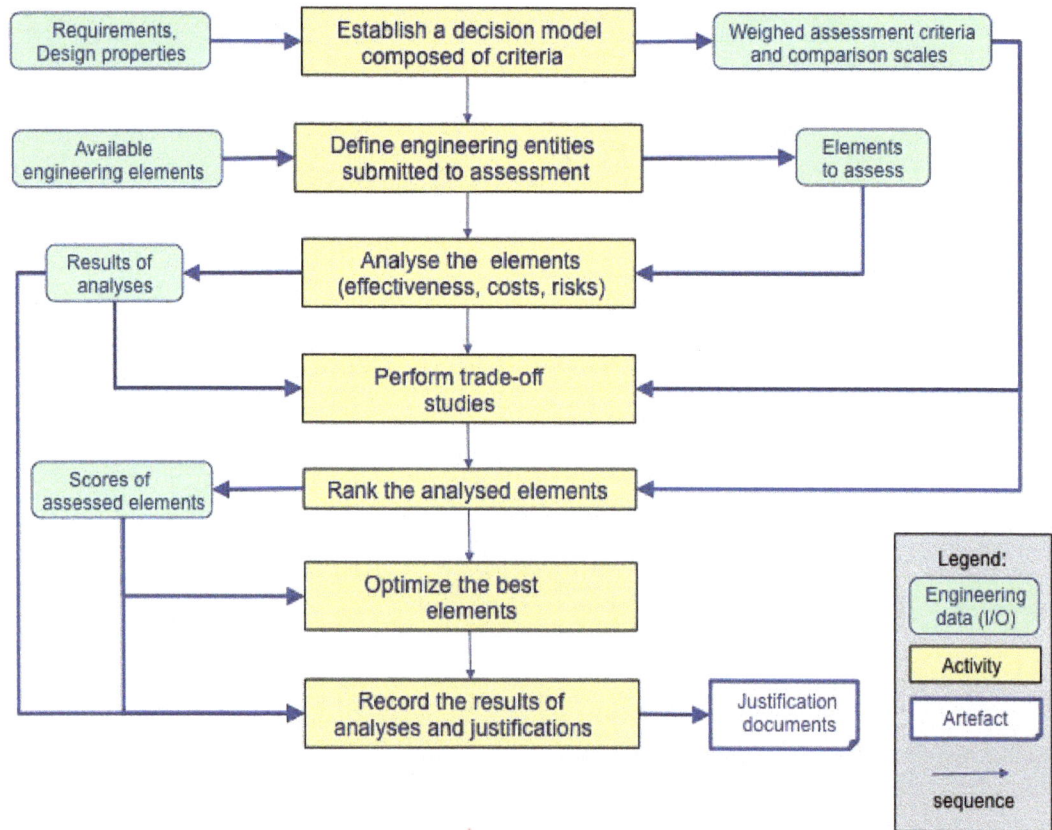

Figure 31 - Activities of the assessment process

Detail of activities

A. Establish a decision model composed of criteria:

1. Define pertinent **assessment criteria** and their limits. These criteria come from the applicable requirements of the system (effectiveness of functions coming from the required services or functions derived from the solution), operational conditions or constraints (efficiency, cost of the solution, technical risks), architectural characteristics or design properties (reliability, maintainability, safety, modularity, simplicity, etc.) the candidate solutions must satisfy. Do not forget to define the limits of criteria; for example, the required minimum efficiency, the maximum cost, minimum design properties values. To do this, input documents, experts' knowledge, stakeholders (customer, users, operators, maintainers, etc.) interviews are used.

2. Quantify the **relative weight** of these criteria. Once the list of criteria is defined from the inputs, the criteria are ranked and weighted. The ranking can be done in a hierarchical tree structure if the criteria are many. In the tree structure, a node will not be decomposed into more than ten criteria - see examples in section 4.1.5. The relative weighting of each criterion is decided on for the proper functioning of the system, for compliance with requirements or by stakeholders. In the latter case, a consensus between the stakeholders is necessary to decide the relative weight between the criteria. In any case, it is recommended to involve several experts to validate the ranking of the criteria. Weighing can be achieved by marking each criterion from 1 to 10 by each expert and then averaging and normalizing the result over the interval,] 0, 1 [.

3. Establish an appropriate **comparison scale** for each selected criterion. The comparison scale depends on the nature of the criterion. This task is explained in sections 4.1.1, 4.1.2, 4.1.3, and 4.1.4.

B. Define candidate solutions or engineering entities submitted to assessment. Depending on the purpose of the evaluation, these elements may be:

- alternative operational or technological concepts
- sets of stakeholder requirements
- sets of system requirements
- candidate logical architectures
- candidate physical architectures
- alternative verification, validation, integrations strategies

Note: In a system vision, the assessment process can be seen as an indicator of **innovation**. Indeed, this process is not only used to evaluate solutions that meet the requirements. Sometimes it is impossible to change the characteristics of the solution and therefore to satisfy stakeholder requirements. At this point, the requirements must be reconsidered. This is one of the reasons why it is important at the beginning of the project to write a document that includes the expression of needs (or stakeholder requirements) that includes a **level of flexibility** for each need or expectation. Assume that stakeholder requirements contain contradictions between two requirements: solution S1 meets Ex1 requirement and does not meet Ex2 requirement, solution S2 meets Ex2 requirement and does not meet Ex1 requirement. If we do not change anything in the requirements or in the solutions, we are in a blocked situation. On the other hand, if we consider the "requirements-solutions" pair, it is likely that a compromise between the two requirements may create a new requirement and thus create a new solution that satisfies this new requirement; we are in a situation of innovation.

C. Analyse candidate solutions or elements submitted to assessment:

1. Perform operational **effectiveness** analyses - Principles and details of this task are described in section 4.1.4.
2. Perform **cost** analyses - Principles and details of this task are described in section 4.1.2.
3. Perform technical **risk** analyses - Principles and details of this task are described in section 4.1.3.

Note: This activity requires representation models and analytical models - see section 4.3.1.

D. Perform trade-off studies for each assessment criterion:

1. Collect the results of above analyses. These are simulation, calculation, and estimates results.
2. Position each engineering element or each candidate solution on the comparison scale of each assessment criterion. If a criteria tree is used, only the leaf criteria are concerned.
3. Present the results in a table or graphics in order to be able to make comparisons.

E. Rank candidate solutions or analysed elements - see section 4.1.6:

1. For each candidate solution or assessed engineering element, calculate a global score. If the number of criteria is important (in the case of multi-criteria models), this global score is calculated by combining the weighed assessment criteria and the marks on the comparison scales of the criteria obtained by the solutions or assessed elements.
2. Compare **scores** as values or in graphics.
3. Perform sensitivity tests on the criteria of the decision model. The choice is said to be robust if a small change in the value of a relative weight does not tilt the choice of solution. In the opposite case, if the choice is fragile, review or modify or add criteria, solutions, analyses, models or rules of choice. Iterate until a robust choice is possible.

F. Optimise the best elements or selected solutions:

1. Try to improve the score of the selected solution by considering the scores obtained by the solutions discarded for certain criteria - see section 4.1.6.
2. Make recommendations for technical decision-makers and calling processes.

G. Record the results of analyses and justifications:

1. Record the results of analyses, arguments for choice or justification, recommendations in the engineering database. These results include assumptions, details of analyses, difficulties encountered, lessons learned, models used, decisions made and their rationales, and any information that will allow the understanding or interpretation of the results.
2. Provide the records to calling processes.

Iterate the activities or tasks of the process when necessary; i.e. when the estimates or calculations are not sufficiently precise or do not allow technical decisions on the choice of engineering elements or a satisfactory solution.

4.2.4 Ontology elements

Elements

The assessment process uses the main engineering meta-data indicated in the Table 5.

ELEMENT	DEFINITION AND ATTRIBUTES (examples)
Assessment criterion	Characteristic to assess and compare engineering elements between them.
	Identifier; Description; Comment; Relative weight; Aggregated weight
Assessment score	Calculated value using a combination of weighted criteria with each other and the marks assigned for each criterion to an assessed engineering element.
	Identifier; Description; Value; Comment
Assessment selection	The element that is the object of assessment. Tree root in a multi-criteria model.
	Identifier; Description; Comment
Cost	Estimated, assessed or calculated value of the amount allocated to an object (e.g. Component, Physical Architecture, Physical interface), or an activity of the life cycle, expressed in a given currency at a given time.
	Identifier; Description; Value; Comment
Design property	A (quantitative) measure or an estimate of an architectural characteristic. It is associated with a system component, a physical interface, a physical architecture. It is a property obtained when defining the components of the system via the allocation of non-functional requirements or obtained by means of techniques (estimation, analysis, study, calculation, simulation) with respect to a specific aspect, or obtained by definition in the case of an existing component.
	Identifier; Description; Comment
Engineering entity	Engineering meta-data: Stakeholder Requirement, System Requirement, Scenario (for logical architecture), Physical Architecture, Component (System element), Physical Interface, etc.
	Identifier; Description; Comment
Rational	A justification action is to demonstrate the relevance of an action (choice, decision, approach, etc.) by providing evidence or written arguments.
	Identifier; Description; Comment (reasons of the relevance of the selected element)
Risk	(Used for technical risk in engineering) An event having a probability of occurrence and a gravity degree on its consequence onto the system mission or on other characteristics. A risk is a measure of the combination of vulnerability and of hazard or threat.
	Identifier; Description; Criticality; Comment; Status

Chapter 4.2. System properties assessment: Process approach

ELEMENT	DEFINITION AND ATTRIBUTES (examples)
System requirement	Statement that identifies an expected characteristic of a product, service, enterprise or process operational, functional, or constraint, which is unambiguous, testable or measurable, and necessary for product, service, enterprise or process acceptability. (Adapted from ISO/IEC 26702 = ISO/IEC 24748-4)
	Identifier; Description; Type; Comment

Table 5 - Main engineering meta-data related to assessment

Relationships

The main relationships between the engineering meta-data are presented in Figure 32.

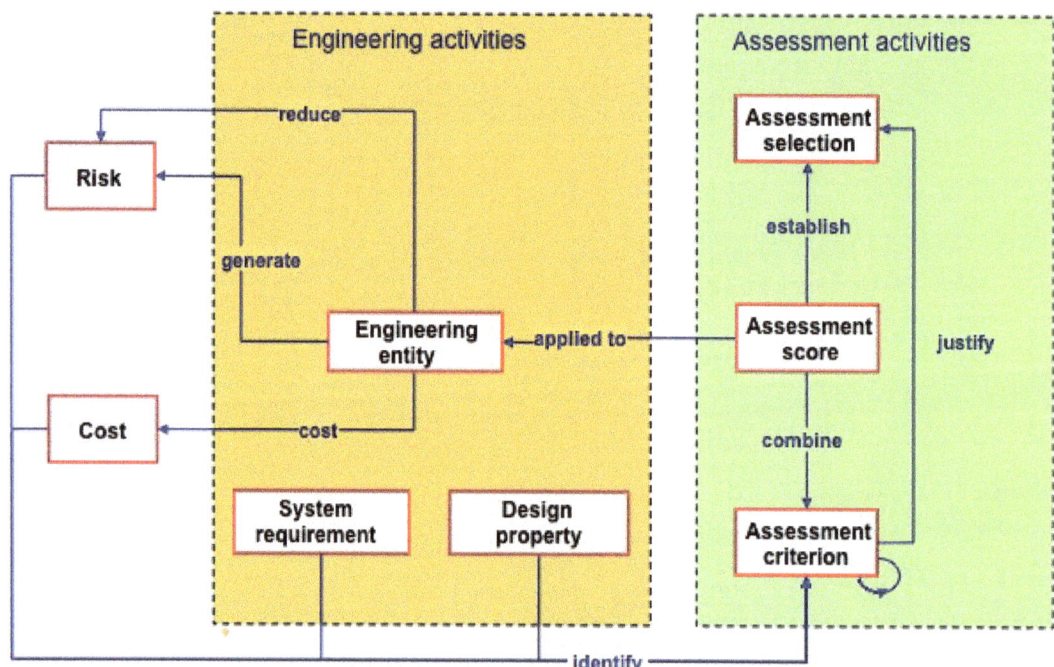

Figure 32 - Relationships between engineering meta-data related to assessment

Utilisation of ontology elements

The meta-data defined above in Table 5 and their relationships in Figure 32 represent the skeleton of the activities and tasks of the Assessment Process. During the execution of tasks of the process the experts in systems engineering have to:

- Create as many as necessary instances of these meta-data: Assessment selection (the subject of evaluation), Assessment criterion (comparison characteristic coming from Requirements or Design properties), Assessment score (global value of comparison);

- Fill in attributes of these instances with values; the generic attributes represent characteristics of a meta-data; particular values for one instance particularise this instance among others;
- Instantiate the generic relationships between meta-data to establish traceability links between instances.

The main relationships to establish during the execution of this process are the following:

- an **Assessment criterion** is **identified by** a **System requirement**, a **Design property**, a **Cost**, a **Risk**
- a **System requirement**, a **Design property**, a **Cost**, a **Risk**, **identifies** an **Assessment criterion**
- an **Assessment score combines Assessment criterion**
- an **Assessment criterion justifies** an **Assessment selection**
- an **Assessment score establish** an **Assessment selection**
- an **Assessment score** is **applied to** an **Engineering entity**
- an **Engineering entity costs** a **Cost**
- an **Engineering entity generates** a **Risk** (possibly)
- an **Engineering entity reduces** a **Risk**

Implemented by software applications such as Data Bases or Repositories or Spreadsheet tools, these relationships enable the generation of Traceability Matrices; these latter are useful for various purposes such as verification, impact analysis (in case of evolutions), justification, technical risks management, and consistency checking.

4.2.5 Verification and validation of evaluations

Main items to be verified and validated during the execution of the assessment process are summarized hereafter:

- The number of assessment criteria is sufficient to differentiate candidate solutions or elements
- The choice of criteria is relevant to the evaluation subject
- The relative weight of the criteria are agreed by stakeholders; a consensus is established
- The comparison scales are consistent between them (minimum and maximum are defined)
- The candidate solutions or elements submitted to evaluation are verified and validated
- Trade-offs of couple "need-solution" are studied; stakeholder requirements are negotiated between acquirer and supplier
- The precision of analysis of assessed properties or characteristics (effectiveness, cost, risk) is sufficient to differentiate candidate solutions or elements
- The effectiveness analysis method is defined and respected
- The cost analysis method is defined and respected

- The risk analysis method is defined and respected
- Global scores are correctly calculated; the same calculation formulas are applied to each candidate solution or element
- Sensitivity tests onto assessment criteria are done in order to obtain a robust selection
- Optimisation studies are tried in order to improve the selected solution
- Recommendations are expressed towards the evaluation applicants or calling processes

4.2.6 Artefacts - documentation

This process essentially generates documents for selection justification or rational. The content of a System Definition Justification Document is presented below. A template extended to the system integration and guidelines are provided in Annex chapter 8.1. This template must be adapted to the type of project and business case.

Content of the System Definition Justification Document (SDJD)

1	INTRODUCTION
1.1	Presentation of the document
1.2	Overview of the system-of-interest
1.3	Documents
1.3.1	Reference Documents
1.3.2	Applicable Documents
1.4	Terminology: definitions and abbreviations
2	JUSTIFICATION OF SELECTED STAKEHOLDER REQUIREMENTS
2.1	Stakeholders
2.2	Rationale for selection of Stakeholder Requirements
3	JUSTIFICATION OF SELECTED SYSTEM REQUIREMENTS
3.1	Rationale for selection of System Requirements
3.2	Traceability between Stakeholder Requirements and System Requirements
4	JUSTIFICATION OF THE LOGICAL ARCHITECTURE
4.1	Rationale for selection of functional, behavioural, temporal architectures
4.2	Traceability between functional elements and System Requirements
5	JUSTIFICATION OF THE PHYSICAL ARCHITECTURE
5.1	Rationale for selection of physical architecture, its components and interfaces
5.2	Traceability between components, functions and System Requirements
6	Mitigation of technical risks

4.3 Practice

This chapter provides explanations, examples and recommendations related to evaluations:

- The applicable models - section 4.3.1, including:
    - Generality about modelling - 4.3.1.1
    - Deterministic models - 4.3.1.2
    - Stochastic models - 4.3.1.3
- Expert opinion - section 4.3.2
- A simple assessment example - section 4.3.3
- Useful practices, recommendations and FAQ - section 4.3.4

4.3.1 Applicable models

4.3.1.1 GENERALITY ABOUT MODELLING

Types of models

We can distinguish physical models, schematic models and mathematical models:

- **Physical models** make it possible to simulate certain technical characteristics of products or physical phenomena; for example: mock-up, vibration table, test bench, CAD tool, virtual environment, pressure vessel, wind tunnel, etc. They are mainly used for technological design, by the relevant professions. These models are not discussed here.

- **Schematic models** are mainly used to simulate the behaviour of a system (for example the FFBD - Functional Flow Block Diagram, State-Charts or State-Machines, State-Transition diagrams, PETRI Nets, etc.), or to represent structural compositions (for example the PBD - Physical Block Diagram, or the IBD - Internal Block Diagram). These models are used and described in Volume 3, Systems Architecture and Design [Faisandier3 2013].

- **Mathematical models** are mainly used to quantify design properties, to calculate dimensions, costs, constraints, etc. Here we limit ourselves to two categories of mathematical models: **deterministic models** and **stochastic models** (also called probabilistic models). The purpose of mathematical models is to transform inputs from the system into outputs using a set of equations or graphs to approximate as much as possible the actual operation of the system. They can be simple (for example an addition), or more complicated (e.g. a probabilistic distribution with several variables).

What level of accuracy?

Pre-project or early project studies use simple models that allow coarse grain assessments, which have the advantage of not being too time and energy consuming, and often sufficient to eliminate unrealistic candidate solutions that are either unfeasible or outside the limits imposed by the context of use, budget constraints or unacceptable risks.

As the project and the maturity of engineering data progress, it is necessary to refine the models and consolidate increasingly precise and coherent data to compare the candidate solutions that have remained in contention. The exercise is all the more complicated as the level of innovation increases.

It is important to use models adapted to the progress of the project: wanting a very high level of accuracy is time and energy consuming. And this is often impossible, because the data do not exist yet! Unfounded assumptions must be avoided.

Modelling is a challenging activity that requires a great deal of technical knowledge, both about the system environment and about the various technologies used. Alone, a single engineer or system architect cannot model a complex system: he must use competent people in the various concerned disciplines.

What data as input?

A model needs input data. Data collection and input variables, as well as model parameters are important tasks: although they are long and sometimes not very rewarding, they are essential for obtaining modelling results of a good quality level. If the input data are not correct, the model will only issue poor results. The more detailed and accurate data the engineer has, the easier the comparative studies.

The input data can come from several origins:

- From systems previously developed, similar in some respects, and from which some of the data can be taken up. This requires that the recording of the conceptualization and experience feedback of previous projects be properly carried out: presence of engineering databases, including details about the environment of use, as well as valid information.
- From simulations results previously carried out, including their environment, conditions and accuracy level.
- From interviews with concerned professional experts.
- Etc.

Limits of models and precaution of use

- A mathematical model gives a mathematical result from mathematical data. It is not a decision-making tool.
- Models can never simulate all the reactions of a system: they only work in a limited domain, on a limited number of variables. When using a model, check that parameters and input data are included in the operating range; otherwise the model provides aberrant results.
- Models must evolve throughout the project:
    ♦ either by modifying the parameters or by adding data when there is a modification or addition of evaluation criteria, functions, objectives and/or derived requirements,
    ♦ or by using new models when those used reach their limits.
- It is recommended to use several models in parallel in order to compare the results of the analyses.
- A model can be adapted to a level N of the system and not adapted to the higher levels of system decomposition. It is therefore essential that the engineer verifies or consolidates the consistency of the different models used in the system tree. One must remember, concerning this point, that the optimization of the system is not the sum of the optimization of the different subsystems.

- It is always necessary to have a critical viewpoint about the models: studying one (or some) properties of the system may hide undesired phenomena, but is likely to develop important undesirable consequences. This is the case for emerging properties.
- The results of a simulation must always be interpreted in relation to their modelling context: tool used, assumptions used, parameters and data entered, and confidence interval of outputs.
- It may be useful to look at models used in previous projects or models available on the market: this avoids "reinventing the wheel", and therefore saves precious time. If the model has been validated against the reality it is supposed to represent, it can be trusted.
- In the case of a purchased model, make sure that the equations used are clearly explained and that the parameters (or even the equations themselves) can be modified by the user: it is essential to control the simulation, and therefore to trust the results.

4.3.1.2 DETERMINISTIC MODELS

Models based on statistics

Statistical models are based on a large number of analyses and previous results. These models require the presence of many data from previous projects and can only be applied to components of the same technology.

For example, at a given time, in the aeronautical domain, the cost C of a transmitter-receiver module was modelled by the formula

$$\ln C = 1.1110 \ln W + 0.436 \ln N$$

where W is the weight of the module and N is the number of modules purchased.

Models based on analogy

The models based on analogy also use data from previous projects: the component studied is approximated to an existing component whose characteristics (cost, reliability, ambience resistance, dimensions, parameters, etc.) are known. Then these characteristics are adjusted on the basis of expert opinions.

Bottom-up models

Bottom-up models require more work: it is necessary to take into account the characteristics of each component and to combine them to obtain an overall result. These models should be reserved for final analyses or innovative technologies, as they require a great deal of time and resources.

Learning curves

Learning curves are used to predict the evolution of a cost (or other characteristic) over time. For example, whenever the number of units produced is multiplied by two, the cost of this unit is reduced by a certain percentage, generally constant. A cost learning curve is expressed in the equation:

$C(n) = C(1) \; n^b$

with $C(n)$ = cost of the nth unit, $C(1)$ = cost of the first unit and b = constant.

Chapter 4.3. System properties assessment: Practice

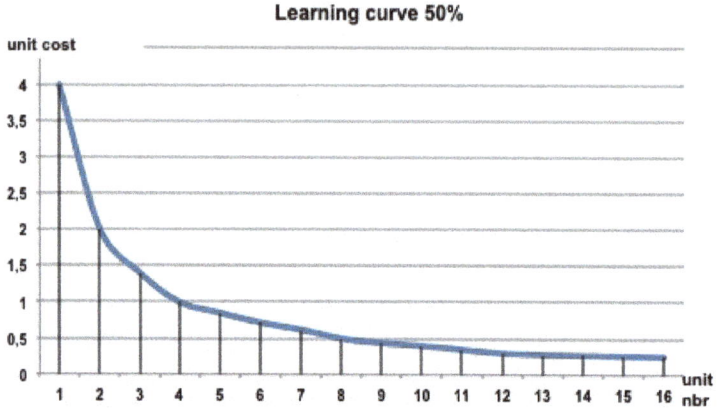

Figure 33 - Example of a cost learning curve

By definition, an LR learning rate of 0.9 (or 90%) means that the cost of the second unit is equal to 90% of that of the first. We then have:

$$b = \ln(LR) / \ln(2) \quad \text{and} \quad LR = C(2n) / C(n)$$

One can take into account either the average of the costs of the produced units (cumulative average learning curve) or the cost of the last produced unit (linear learning curve).

Beta curves

Beta curves are used to model the distribution of a variable over time. For example, at the percentage of project completion time t, the amount spent can be written as:

$$C = 10t^2 (1-t)^2 (A + Bt) + t^4 (5 - 4t) \quad \text{with} \quad 0 \le t \le 1 \text{ and } 0 \le A + B \le 1$$

Parameters A and B are adjusted according to the experience of previous similar projects. Generally, one uses a profile 50% (A = 0 and B = 1) or 60% (A = 0.32 and B = 0.68).

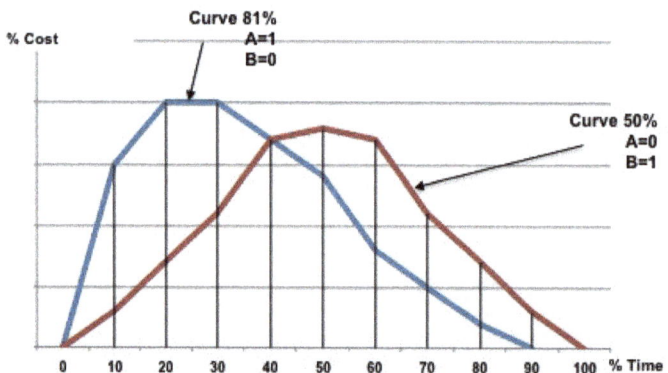

Figure 34 - Example of a beta curve with 50% and 81% profiles

4.3.1.3 STOCHASTIC MODELS

The probability calculations make it possible to classify the candidate solutions when the number of evaluation criteria is limited, and the combination of the criteria is simple. Ensure that the sum of the probabilities of the various criteria is equal to 1 for each node.

A simple example of a calculation of probabilities to go from city V to city U, taking more or less probable events into account:

Candidate solutions	Event 1	Probability	Event 2	Probability	Global probability
Highway	-	0.8	-	-	0.8
	accident	0.01	total blocking	0.8	0.008
			blocking <1h	0.2	0.002
	traffic jam	0.19	blocking <1h	0.4	0.076
			blocking >1h	0.6	0.114
Standard road	-	0.7	-		0.7
	accident	0.02	total blocking	0.3	0.006
			blocking <1h	0.7	0.014
	traffic jam	0.28	blocking <1h	0.7	0.196
			blocking >1h	0.3	0.084

Note: The probabilities shown in this table are invented and do not correspond to any reality.

From the previous table we can deduce the following table, which gives indications to help in making a decision:

	Highway	Standard road
No event	0.8	0.7
Delay < 1 h	0.078	0.21
Delay > 1 h	0.114	0.084
No arrival	0.008	0.006

Reminder of simple probability calculations in different combinations of events:

- Probability that A is realized when B has already been realized:
  - ◆ P = Pa . Pb
- Probability that two independent events A and B are realized:
  - ◆ P = Pa + Pb - (Pa . Pb)
- Probability that three independent events A, B and C are realized:
  - ◆ P = Pa + Pb + Pc - Pa . Pb - Pb . Pc - Pc . Pa + (Pa . Pb . Pc)
- **For more complex probability calculations**, there are computational tools for combining distribution curves. See also the Monte-Carlo simulation.

4.3.2 Experts opinion and interviews

When a criterion or value cannot be determined precisely, or the subjective aspect predominates, experts' opinion or interviews can be used to approach a realistic value.

The operation is carried out in four stages:

- Selection of persons to be interviewed
- Writing a questionnaire
- Interviews
- Analysis of results

Only small targeted surveys are being discussed here; for public inquiries, call on professionals.

Selection of persons to be interviewed

It is necessary to choose persons competent for the field in question.

For example, in order to establish assessment criteria whose objective is to classify the candidate solutions, the client-user and the decision-makers of the company are consulted.

On the other hand, in order to establish the risk of failure of an engine or to give a mark of maintainability to different candidate solutions, specialists are selected.

Questionnaire

The more precise the questionnaire is, the easier its operation. On the other hand, an overly closed questionnaire may leave out important points. To avoid this pitfall:

- have the questionnaire written by several people with different approaches,
- do not hesitate to change the questionnaire during the survey, even if it means reviewing the people already interviewed.

When the questionnaire is about classification or scoring, two methods can be used:

- Ask the interviewee to have an overall view and rank the items in order of preference, or give them a mark; the marking method is preferable, provided that the rating scale is well calibrated and accurate.
- Present the elements in pairs and ask each time which element seems most important, by assessing this relative importance; the method is longer but more precise.

Interview

Questionnaires can be sent by mail, which is easy and allows the opinion of many people; but this is not very effective.

It is best to have a thorough discussion with a limited number of people who give a more complete and accurate opinion.

Analysis of results

It is preferable to have the results of the questionnaires analysed by several different people. They can compare their impressions or opinions, until they agree on a ranking.

Comparison technique 2 to 2

Let us examine the example that consists of classifying 8 criteria, denoted from A to H.

One begins by having a matrix representing the relative weight of each criterion compared with all the others. For example, the interviewee compares two criteria A and D, with the following scale of importance:

1	Same importance
3	Slightly higher importance
5	Higher importance
7	Significantly higher importance
9	Very high importance (the other becomes negligible)
2-4-6-8	Intermediate values between

For example, if criterion A is considered to be very much greater in importance to D, we note 7 in box A / D and, conversely, 1 / 7 = 0.14 in box D / A

The fully completed comparison matrix is as follows:

	A	B	C	D	E	F	G	H	m	8Vm	W
A	1.00	5.00	3.00	7.00	6.00	6.00	0.33	0.25	315	2.05	0.175
B	0.20	1.00	0.33	5.00	3.00	3.00	0.20	0.14	0.09	0.74	0.063
C	0.33	3.00	1.00	6.00	3.00	4.00	6.00	0.20	86.40	1.75	0.149
D	0.14	0.20	0.17	1.00	0.33	0.25	0.14	0.13	0.00	0.23	0.019
E	0.17	0.33	0.33	3.00	1.00	0.50	0.20	0.17	0.00	0.42	0.036
F	0.17	0.33	0.25	4.00	2.00	1.00	0.20	0.17	0.00	0.50	0.042
G	3.00	5.00	0.17	7.00	5.00	5.00	1.00	0.50	218.75	1.96	0.167
H	4.00	7.00	5.00	8.00	6.00	6.00	2.00	1.00	80640	4.11	0.350
										11.74	

The weight of each criterion is calculated as follows:
- Multiply all the marks of the line to obtain m
  - ♦ Here, for A: $m = 1 \times 5 \times 3 \times 7 \times 6 \times 6 \times 0.33 \times 0.25 = 315$
- Extract the root Nth of the result
  - ♦ Here, there are 8 criteria, therefore the root 8th (8Vm) is calculated

Chapter 4.3. System properties assessment: Practice

- Transform the result to 1 dividing the root by the sum of roots
  - *Here, for A: 2.05 / 11.74 = 0.175   so the weight of criterion A is 0.175*

Then check the sensitivity of the result as follows:
- Sum each column
  - *Here, for A: Tot = 9.010*
- Multiply this sum (Tot) by the weight W of the corresponding criterion
  - *Here, for A: 9.010 x 0.175 = 1.575*

This is shown in the last two lines of the matrix:

	A	B	C	D	E	F	G	H	m	8Vm	W
A	1.00	5.00	3.00	7.00	6.00	6.00	0.33	0.25	315	2.05	0.175
B	0.20	1.00	0.33	5.00	3.00	3.00	0.20	0.14	0.09	0.74	0.063
C	0.33	3.00	1.00	6.00	3.00	4.00	6.00	0.20	86.40	1.75	0.149
D	0.14	0.20	0.17	1.00	0.33	0.25	0.14	0.13	0.00	0.23	0.019
E	0.17	0.33	0.33	3.00	1.00	0.50	0.20	0.17	0.00	0.42	0.036
F	0.17	0.33	0.25	4.00	2.00	1.00	0.20	0.17	0.00	0.50	0.042
G	3.00	5.00	0.17	7.00	5.00	5.00	1.00	0.50	218.75	1.96	0.167
H	4.00	7.00	5.00	8.00	6.00	6.00	2.00	1.00	80640	4.11	0.350
Tot	9.010	21.867	10.250	41.000	26.333	25.750	10.076	2.551		11.74	
Tot*W	1.575	1.370	1.524	0.793	0.937	1.089	1.683	0.892	9.863		

- Add the results of the last line to find Imax.
  - *Here: Imax = 9.863*
- Calculate SI = (Imax - n) / (n - 1), where n is the number of criteria (=8)
  - *Here: SI = (9.863 - 8) / 7 = 0.266*
- Calculate the sensitivity rate SR = SI / F(n), with the following values of F(n):

n	1	2	3	4	5	6	7	8	9	10
F(n)	0	0	0.58	0.90	1.12	1.24	1.32	1.41	1.45	1.49

  - *Here SR = 0.266 / 1.41 = 0.189 = 18.9 %*

If the sensitivity rate is less than 10%, the ranking is consistent. A value of sensitivity can be accepted up to 20%, but beyond, it is necessary to resume the classification.

This method can be used for any type of classification: classification of criteria, characteristics, candidate solutions, etc.

4.3.3 Assessment example

A simple example to understand: selection of a trip

- Criteria and their relative weight:

Criteria	Weights
Cost	0.1
Speed	0.5
Safety	0.4
Total	1.0

- Three candidate solutions (marks on 10 provided by calculation or expert opinion):

Criteria	Trip 1	Trip 2	Trip 3
Cost	2	7	5
Speed	4	2	1
Safety	6	4	7

- By multiplying the mark of each cell of the table by the weight, the following table is obtained:

Criteria	Trip 1	Trip 2	Trip 3
Cost	0.2	0.7	0.5
Speed	2	1	0.5
Safety	2.4	1.6	2.8
Score	**4.6**	**3.3**	**3.8**

- For the selected criteria, the best solution is trip 1 (the higher score).

4.3.4 Recommendations and frequently asked questions

Some pitfalls often encountered are indicated in Table 6:

PITFALL	DESCRIPTION
An analytical model is not a decision tool	An analytical model provides analytical results from analytical data. It must be seen as an aid and not as a decision-making tool.
Models and system decomposition levels	Models used at the decomposition level n of a system are mostly incompatible with higher-level models that use data from the lower level. It is essential that the system engineer checks the consistency of the different models used.
Global and local optimization	The global optimization of the system is not the sum of its optimized subsystems.

Table 6 - Pitfalls about assessment

Some good practices are indicated in Table 7:

PRACTICE	DESCRIPTION
Respect the scope of the model	Models can never simulate all the behaviours / reactions of a system: they only work in a limited domain with a limited number of variables. Parameters and input data must be included in the scope.
Evolve models	The models must evolve during the project: by modifying the parameters, by entering new data; by the use of new tools when those used reach their limits.
Use several types of models	It is recommended to use several types of models simultaneously in order to compare the results and/or to take into account another aspect of the system.
Keep consistency with context elements	The results of a simulation must always be accompanied by a description of the modelling context: tool used, assumptions used, parameters and data entered, variance of results, etc.

Table 7 - Good practices about assessment

FAQ

What does the term "analysis" mean?

The process that we have named here System Properties Assessment is named System Analysis Process for systems engineering experts.

The term "system analysis" is the contraction of "analysis of the existing / under development system to assess its properties". In this book, we wanted to be directly explicit in the title of the process.

As explained elsewhere, the term "analysis" refers to something that exists, or one can have a good representation of its properties. This is why this process relies heavily on models of system properties.

What is the utility of keeping justifications or rationales?

In the early stages of the development of a system, numerous studies are carried out: feasibility studies, technical choices, architectures studies, technological assessment, etc.

All the information gathered must be kept in order to construct the technical memory of the system and the related project, and make it possible to know a few years later why such a choice was made. This information constitutes the theoretical justification and is useful from a contractual viewpoint. It is essential to retain and organize this information, especially as the staffs responsible for these studies are no longer accessible.

5 VERIFICATION AND VALIDATION OF THE SYSTEM

This chapter presents all the means necessary to carry out the verification and validation activities during the life cycle of the system. It emphasizes the development of the system, i.e. its engineering and its integration.

This chapter presents:

- The verification and validation concepts, principles and approaches - Chapter 5.1, which includes:
  - The definitions - section 5.1.1
  - The notion of verification and validation action - section 5.1.2
  - The verification and validation techniques or methods - section 5.1.3
  - An overview of verifications and validations - section 5.1.4
  - The verification and validation strategy - section 5.1.5
- The process approach - Chapter 5.2, which includes:
  - The location of processes - sections 5.2.1.1 and 5.2.2.1
  - The purpose, inputs and outputs of the processes - sections 5.2.1.2 and 5.2.2.2
  - The description of activities of the processes - sections 5.2.1.3 and 5.2.2.3
  - The ontology elements - sections 5.2.1.4 and 5.2.2.4
  - The artefacts generated by the processes - sections 5.2.1.5 and 5.2.2.5
- The practice - Chapter 5.3, which includes:
  - The justification traceability and matrix - section 5.3.1
  - Examples of verification and validation actions - section 5.3.1
  - The technique of inspection - section 5.3.2
  - The technique of review - section 5.3.3
  - The technique of test - section 5.3.4
  - The organisation of activities and responsibility - section 5.3.5
  - Recommendations and FAQ - section 5.3.6

5.1 Concepts, principles, approaches

5.1.1 Definitions

Reading Chapter 3.5 is a prerequisite for understanding the present chapter; it is therefore strongly recommended.

> **Why verify?**

Verification is a necessity, because the human being in his actions makes mistakes. To do everything perfectly from the first attempt is quite simply impossible. Error is an integral part of thinking and human action; an individual trained in a type of operation makes at least 10^{-3} errors per hour; this figure should be reduced most of the time to 10^{-2}, or even 10^{-1} depending on the conditions of the operational environment and the personal state of the operator (mental load, stress, emotion, attitude, etc.). There is therefore no dishonour in integrating error search into the execution or results of an engineering process; on the contrary, it is an indicator of competence.

From a psychological viewpoint, it is important not to treat error as a *punishable fault*, because such an attitude blocks creativity or generates fearful reactions to any initiative, a fortiori within the context of a project. On the other hand, overlooking the search for potential errors in a system, or in the execution of engineering processes makes the project take significant risks.

Therefore, any initial system (product, service or organization) conceived by human beings contains defects and/or inadequacies. An initial system is the first version of the system; the subsequent versions or evolutions correct the defects and tend, in general, to improve the functioning of the system. The verifications are therefore **normal** and **indispensable**.

The rationale for the verifications lies in the following two questions:

- Is the product, service or organization obtained through the execution of the engineering, realization and integration processes, the **expected** one?

- When using the product, service, or organization, do residual defects and errors **impact** on itself (mission, functionality, performance, safety, etc.) and its environment or context (safety of persons and property, etc.)?

The stakes that must guide the verifications to be carried out are:

- technical, in particular with regard to aspects of
  - ♦ conformance (does the product, service or organization correspond to the needs?)
  - ♦ operability (ability to use the product, service or organization)
- social, for
  - ♦ safety of persons and goods
  - ♦ quality of the user service
- economical, in particular with regard to aspects of
  - ♦ supply market
  - ♦ industrial competition
  - ♦ costs and deadlines

Chapter 5.1. Verification and validation of the system: Concepts and principles

Definition of verification

Verification is an activity intended to determine whether the system (product, service, organization) meets the expected **design properties** or characteristics. Verification properties are general (independent of requirements) or specific (deducted from requirements).

Discussion

A **verification action** consists of identifying the defects introduced during the transformation of the inputs into outputs of a task, of an activity or of a process, with the aim of eliminating them. It serves to determine that the transformation was **well made**. It is based on tangible evidence, that is to say on information whose veracity can be demonstrated, based on facts obtained by observation, measurement, testing, analysis, calculation, etc.

The following examples of application clarify the notion of verification:

- Verify a product consists of comparing the properties of the realized product against its expected design and implementation properties, to detect defects. These properties are either general and independent of the requirements (they come from the state of the art, decisions of the architect or the designer), or specific and deduced from the applicable technical requirements.

- Verify a document is to show that it respects the rules of writing and elaboration related to the task that produced it.

- Verify a process consists of comparing the performance of each process activity against a procedure, rule, standard, or established method.

The verification thus applies to every activity and result of this activity, in particular in engineering.

To verify is to determine that one has *well* made the system.

The standard ISO 9000:2015 gives the following definition of the term verification:

Confirmation, through the provision of objective evidence, that specified requirements have been fulfilled.

> *The objective evidence needed for verification can be the result of an inspection or of other forms of determination such as performing alternative calculations or reviewing documents.*
>
> *Objective evidence* means data (facts about an object) demonstrating the existence or veracity of something. Objective evidences are obtained by observation, measurement, test or other means. The term *specified requirements* should not be confused with specific requirements. The term *specified requirements* in fact means expected characteristics or design results.

The standard ISO/IEC/IEEE 15288:2015 [ISO 15288] gives the following definition, with a note:

Confirmation, through the provision of objective evidence, that specified requirements have been fulfilled.

Chapter 5.1. Verification and validation of the system: Concepts and principles

Note 1 to entry: Verification is a set of activities that compares a system or system element against the required characteristics. This includes, but is not limited to, specified requirements, design description and the system itself. The system was built right.

These definitions do not take into account that the verification process is a generic process and must be instantiated according to the type of object to be verified (system, product, document, activity, process, etc.).

Definition of validation

> Validation is an activity intended to gain **confidence** in the ability of the system (product, service, organization) to achieve its mission under specified conditions of use.

Discussion

A **validation action** consists of observing that a realized object (equipment, software, service, product, system, procedure, document, requirement, etc.) satisfies its definition reference. It is based on tangible evidences, that is to say on information whose veracity can be demonstrated, based on facts obtained by observation, measurement, testing, analysis, calculation, etc. In order for confidence to be gained, it is necessary that the definition reference of the object is not questionable, therefore it is itself verified, validated and accepted by the stakeholders.

The following examples of application clarify the notion of validation:

- <u>Validate a document</u> consists of determining that the document is able to fulfil its mission, that is to say communicating to the readers the expected information, in a structure of the information and in a language accessible to these readers.

- <u>Validate a product</u> consists of determining that the product meets its technical requirements and fulfils its functions. Because the technical requirements applicable to the system (system requirements) come largely from stakeholder requirements, a validated product satisfies the stakeholder requirements.

- <u>Validate a process</u> consists of determining that the outcome of the process activities conforms with the objectives assigned to that process.

- <u>The final validation</u> of a system (product, service, organization) is the demonstration of its ability to achieve its mission in an environment representative of operational functioning, especially in borderline or degraded cases. Final validation usually requires validation and verification actions at earlier stages.

The validation thus applies to every activity and results of this activity, in particular in engineering.

> To validate is to determine that one has made the *right* system.

The standard ISO 9000:2015 gives the following definition of the term validation:

Confirmation, through the provision of objective evidence, that the requirements for a specific intended use or application have been fulfilled.

> *The objective evidence needed for a validation is the result of a test or of other forms of determination such as performing alternative calculations or reviewing documents.*

The standard ISO/IEC/IEEE 15288:2015 gives the following definition, with a note:

Confirmation, through the provision of objective evidence, that the requirements for a specific intended use or application have been fulfilled.

Note 1 to entry: A system is able to accomplish its intended use, goals and objectives (i.e., meet stakeholder requirements) in the intended operational environment. The right system was built.

These definitions do not take into account that the validation process is a generic process and must be instantiated according to the type of object to be validated (system, product, document, activity, process, etc.).

5.1.2 Notion of Verification Action and Validation Action

Definitions

Any **verification action** applied to an object (function, product, service, organization, document, requirement, property, component, interface, etc.) assumes:

- to define the action executed on this object,
- the existence of an established reference,
- to define the expected result of this action,
- to acquire a result,
- to compare this result to the expected result,
- to deduce the conformance (scope of the verification).

Figure 35 summarizes the necessary elements for defining and carrying out a verification action.

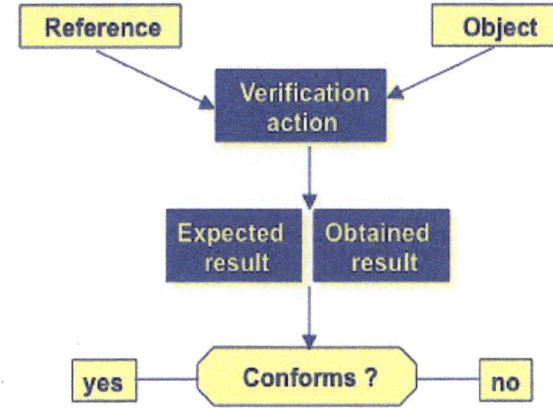

Figure 35 - Necessary elements for a verification action

The comparison between the expected result and the obtained result is binary: there is conformance, or not. When there is uncertainty about conformance, there is ambiguity about the reference (for example, the requirement); the typical example is the performance expressed without any acceptance limit (threshold below or above the performance is declared not reached).

A verification action is carried out using a technique; for example, analysis, inspection, test, review - see section 5.1.3.

Chapter 5.1. Verification and validation of the system: Concepts and principles

Any **validation action** applied to an object (function, product, service, organization, document, requirement, property, component, interface, etc.) assumes:

- to define the action executed on this object,
- the existence of an established reference,
- to define the expected result of this action,
- to acquire a result,
- to compare this result to the expected result,
- to deduce a level of conformance (difference between the obtained result and the expected result),
- to decide on the acceptability of this conformance (scope of validation),
- to decide on the relevance or optimization of the object in its context of use (scope of the justification - see Note).

Figure 36 summarizes the necessary elements for defining and carrying out a validation action.

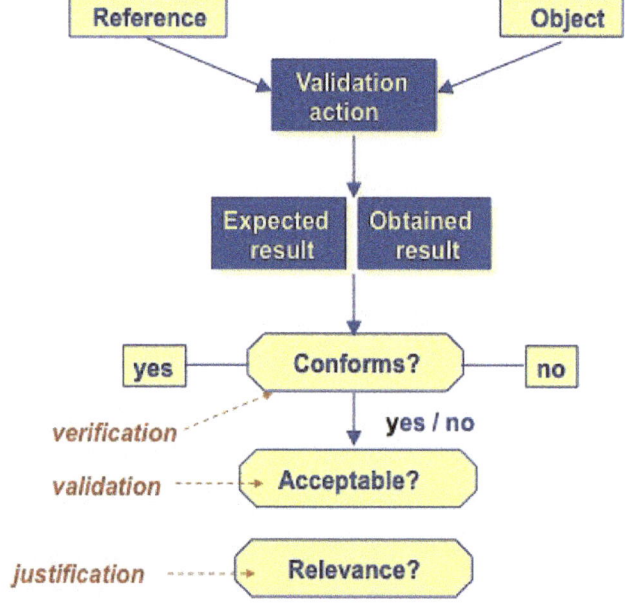

Figure 36 - Necessary elements for a validation action

In some cases, if conformance cannot be ruled out, a decision can be made against a set of criteria, in order to accept, or not, the divergence (e.g., profile of curves); if the result is accepted, there may be *contingency* (reserve); otherwise, correction, repair or remake must take place.

A validation action is carried out using a technique; for example, analysis, inspection, test, review - see section 5.1.3.

Chapter 5.1. Verification and validation of the system: Concepts and principles

Note: The justification, discussed here, is the last step of validation to keep a written record of the relevance of the acceptability, or not, of the discrepancy.

There is another type of justification, that of choice made, subject treated in the system properties assessment - Chapter 4. These two types of justification are independent, but they are mentioned in the System Justification Document - see Annex chapter 8.1.

Difference between verification and validation

The difference between verification and validation may seem rather subtle for the neophyte in terms of verification - validation. For the expert in the field, it is fundamental. A few examples of empirical reasoning make it possible to understand this difference.

Example of a chair:

> For a chair, a verification action consists, for example, in measuring the length of its feet to check that their length conforms with that of the manufacturing plan.
>
> Another verification action may be that the grade of metal conforms with the material imposed by a requirement.
>
> A validation action consists of determining that the chair is able to fulfil its mission as chair; that is to say that an individual can sit down and be comfortable eating, writing, taking a rest or exercising a professional activity.

In this example, there are differences:

- Verification actions give rise to binary responses: the length of the feet is in line with the plan's dimension or not (tolerance interval taken into account). While the validation action may require questions: is the measured height comfortable for the person being considered? Of what person (measurements) is it?

- The verifications concern an element (the foot), whereas the validation concerns the whole (the chair) and considers this whole in its entirety.

- Any validation must be carried out by prior verification actions. If the study has defined that the chair should be 50cm high to be comfortable, it will necessarily have to measure this height to validate that the manufactured chair is comfortable.

Example of building a wall:

> What does the expression of needs to mount an upright wall contain? It is not specified that the wall must be vertical ... it goes without saying! However, it should be checked at the time of delivery. But in any case, the wall will never be exactly vertical. So what difference in verticality is acceptable? It depends on the use and location of the wall.

Example of a virtual object:

In the example of the chair, verification and validation apply to a physical object; virtual objects (plans, documents, processes, etc.) follow the same approach.

> Let's take the example of the following effectiveness requirement: "the vehicle runs at a maximum of 180 km / h with an uncertainty of $\pm$ 2 km / h".

The verification of the requirement consists of determining that it is written in the correct English language (syntax and semantics), that it is numbered, that it is not redundant with another requirement, etc. Whereas to validate the requirement, we ask: is the choice of 180 km / h relevant to the missions that the vehicle must perform? If the vehicle is a road car, the answer will probably be yes; if the vehicle is a public works vehicle, the answer will certainly be no.

The distinction between verification and validation can be formulated as follows:

The verification of an object

- is defined relative to the design properties of this object,
- requires a reference (standard, state of the art rules, patterns, etc.),
- is totally objective,
- can be carried out by the developer of the object,
- must be carried out before any validation action.

The validation of an object

- is defined in relation to the requirements and expectations expressed with respect to this object,
- requires a reference (technical requirements and/or stakeholder requirements),
- involves prior verification actions,
- often requires a value judgment,
- must be carried out by an expert who is not the developer of the object.

Synthesis

To conclude this section, here are some expressions that summarize the notions of verification and validation, as well as their difference or complementarity. These expressions make it possible to make the link with the following sections and with the previous chapter.

- Developing a system without care about verification and validation leads to a random result.
- **WELL** developing a system means that we follow defined processes and they are respected. This is mainly the purpose of the **verification** process - see section 5.2.1.
- Developing the **RIGHT** system means that the system satisfies its technical requirements and stakeholder requirements. This is mainly the purpose of the **validation** process - see section 5.2.2.
- **WELL** developing the **RIGHT** system requires a verification and validation strategy to determine **conformance** and gain **confidence** - see section 5.1.5.
- Developing the **OPTIMAL** system requires the best compromise between all requirements and design properties. This is the purpose of the **assessment** process - see the system properties assessment process, section 4.1.6.
- The justification, or proof, of the system is composed of all the results of the execution of the assessment, verification and validation processes.

5.1.3 Verification and validation methods and techniques

Theoretical justification and experimental justification

The last point in the previous section mentions that the justification activity encompasses evaluation, verification and validation activities. The justifications, or proofs, can be obtained **theoretically** or **experimentally**.

The theoretical justification refers to documents and models. It is obtained, long before the system (product, service, organization) exists, through analysis, inspection or simulation techniques. It does not require significant investment or infrastructure; it detects many defects, inadequacies, inconsistencies or omissions. Nevertheless, it is often difficult to apprehend, because it is based on virtual elements.

The experimental justification concerns realized objects, a product, mock-ups or demonstrators. It is obtained after completion or during integration by inspection, testing or demonstration techniques. It is expensive because it requires real equipment (for example, test benches with real or simulated elements), a more or less complex environment representing real use. It is intellectually easier to apprehend than the theoretical justification.

These two types of justification are complementary and are considered in the development of verification and validation strategies. Some proofs are not acquired through verifications, but through theoretical studies or effectiveness analyses (e.g., mathematical or probabilistic calculations) in evaluations using models - see section 4.1.4.

Static and dynamic implementation

The justifications, or proofs, are obtained by various verification and validation techniques that can be classified in **static implementation** and **dynamic implementation**. Figure 37 shows the relative position of the main verification and validation techniques between static / dynamic implementation and theoretical / experimental justification.

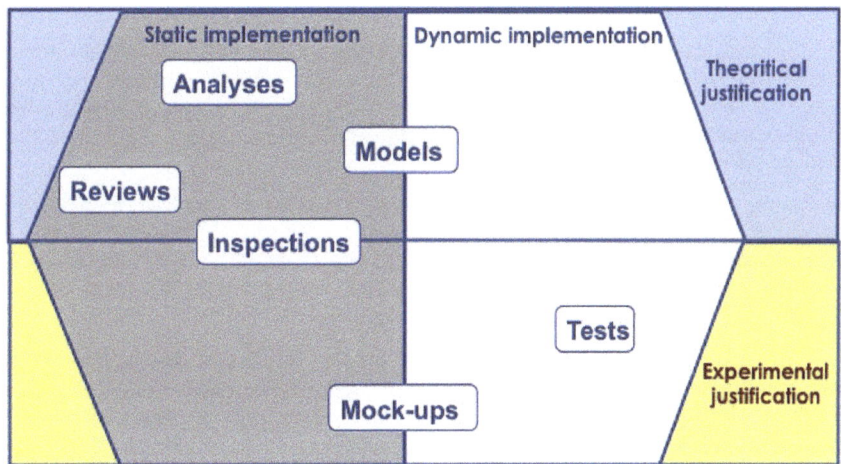

Figure 37 - Position of techniques: static / dynamic and theoretical / experimental

Static implementations do not require the functioning of the system. The object studied is static, whether it is a document, a model, a mock-up, a component, etc. This approach is used to verify and validate documents and models and to analyse the architectural characteristics or design properties of the system, such as technical feasibility, reliability, cost, safety, modularity, reusability, etc.

Dynamic implementations require the functioning of the product, service or organization, whether it is represented by its physical realization, mock-ups or models. This approach is used to put the product, service, organization in conditions close to its operational use and to analyse its behaviour through its functions, nominal and degraded operating scenarios, states changes, etc.

Discussion

Here are some **recommendations** from experience on many projects, for efficiency purpose; i.e. to detect the maximum number of defects and to obtain a good level of confidence in the developed system. Indeed, as we shall see later, it is impossible to obtain a comprehensive set of justifications or proofs for a complex system within a reasonable time in relation to the duration of the development project.

- Begin justifications, verifications and validations **as early as possible** in system development to detect engineering errors, and quickly gain confidence in the future product, service or organization. Many defects can be identified before any physical realization.

- Emphasize the theoretical justification and **limit the experimental justification** to dynamic aspects and insufficiently controlled characteristics or properties.

- The shift from analysis of the results of experimental justifications (based on tests) to the theoretical understanding of the system (based on models) requires good **feedback** bases.

- Take the time to do the actions in the above order, and correctly record the studies, technical choices, analyses results, in order to **retain the lessons learned** thanks to the justifications.

> What seems to be a waste of time in the short term ... is actually a **saving of time** in the medium and long term.

To demonstrate conformance to a technical requirement, the first reaction of neophytes in terms of verification and validation is often to do a test. But a test requires that the system exists (at least in the form of a prototype), implement an appropriate tool set (often expensive and which must itself be validated), know or be able to develop a test representative of reality. While a "paper" study can be much more effective and easy to implement.

For example, verifying a reliability requirement dynamically requires measuring reliability on a demonstrator or even on the actual product. There is a good chance that it will be necessary to wait until the end of the life of the system to obtain such a measure; unless accelerated and often-costly mechanisms of aging are used. It is for this reason that attempts are made to create dysfunction models to assess reliability (predictive failure rate).

In the early phases of the system development project, numerous studies are carried out: feasibility studies, architectural studies, evaluations and comparative studies, technical decisions, etc. All the information collected must be recorded and perpetuated because they contribute to the technical memory of the project. This memory makes it possible to know, a few years later, why a choice was made. This information constitutes the theoretical justification and is therefore useful from a contractual viewpoint. It is essential not to lose this information and to organise it, especially since the personnel, the author of these studies, is ephemeral to the project.

Chapter 5.1. Verification and validation of the system: Concepts and principles

The continuation of this chapter describes the most widely used verification and validation techniques or methods. Whether for verification or validation, the techniques or methods are the same. Only the goals are different: verification is used to detect defects, validation is used to gain confidence - see section 5.1.1.

Analysis techniques: generality

The **analyses** deal with characteristics or properties elaborated throughout the development of a system and its constituent elements, both logical and physical. They use existing engineering data (via models, lists or documents) or concrete elements such as mock-ups or prototypes. They are carried out in different ways:

- by thinking or logical reasoning: deduction, induction, inverse, absurd,
- by mathematical calculation: propositional calculus, predicate calculus, probabilistic computation, equations, etc.
- by analogy, using the results of previous experiments, or by bibliographic analysis, etc.

The analysis techniques are carried out mainly during the upstream activities; feasibility study, requirements engineering, definition of architectures, definition of design of components; but also, in any other activity when a theory or model applies. They do not require heavy technical means.

Practical means are, for example:

- grids of choice: for allocations of functions to components, for criteria of choice of architectures, for choices of algorithms, methods, languages, materials, etc.,
- properties matrices, traceability matrices,
- calculation notes,
- probabilistic estimates for safety,
- power, mass, consumption status, etc.

Analysis technique: formal proof

A **formal proof** is a series of elementary stages of reasoning, called *inferences* (implication by deduction or induction), making it possible to carry out a demonstration. It is useful for demonstrating the validity of mathematical or scientific properties; or conversely to refute properties (e.g. to demonstrate that undesired events cannot occur). It can use a formal language to express certain properties or algorithms.

Note: An inference is an intellectual operation by which one passes from one or more assertions (propositions asserted as true) to a new assertion that is the conclusion. Inference can be run by deduction (from cause to effect, from principle to consequences, from general to particular) or by induction (generalization of a reasoning established from singular cases) or by abduction (introduction of a rule as a hypothesis to continue or stop the reasoning).

Examples of usage of formal proof technique:

- prove a characteristic: demonstrate that the expected characteristic is relevant;
- prove the validity of a set of characteristics: demonstrate that the characteristics are realistic, and identify those that could lead to behaviours to avoid;

- prove that a requirement expression stage is correct: demonstrate the consistency and exhaustiveness of the requirements expressed;
- prove that a step in the definition of an architecture is correct: demonstrate that the requirements of a subsystem, obtained by allocation studies in the architecture activities of the parent system, satisfy the requirements of the parent system;
- prove that a piece of software is correct in relation to its requirements: demonstrate that, for defined inputs, the piece of software determines and delivers the expected results.

Note: For software, we talk about *proof of program*. Proof of program requires:
- a formal language to express the properties of the piece of software,
- a programming language for which a mechanism of proof exists (a tool that verifies that any step of the proof corresponds to an inference rule).

It is not possible to prove a piece of software in its entirety, but this technique can be used for some of its parts. A partial proof cannot guarantee that a piece of software is mathematically correct, but it verifies that it has certain properties deemed critical for safety and expressed by invariants. Partial program proof demonstrates that the piece of software does not react dangerously when the conditions that may lead to catastrophic failures are met. For example, this technique is used for safe software such as emergency braking in rail transport.

Analysis technique: complexity evaluation

We can assume that the number of potential defects in a system is directly correlated to its complexity. The more complex a system is, the more difficult it is to maintain and test, and the greater the risk of residual defects remains. **Complexity measures** are indicators of the difficulty of understanding the architecture and the difficulty of testing the system (dynamic verification and validation).

Some examples of indicators to assess the complexity of a system:
- number of components,
- number of interfaces between these components,
- frequency of exchanges between components,
- interactions between components that may result in emerging properties,
- events combinations,
- necessary time to describe functioning,
- readability of schemas.

Evaluations along the project make it possible to detect the aberrant points of complexity and to reorient the architecture works.

Inspection technique

Inspection is a static verification technique. It relates to the contents of the documentation (document, plan, schema, model, diagram, listing) and/or the realized system in an inert / passive state (without working). The inspection finds that the documentation or the realized system complies with the pre-established rules (standards, plans, procedures). Findings of non-compliance provide the basis for deciding whether or not corrections are made; any corrections are the responsibility of the writers or developers.

The technique uses checklists specific to the object or subject inspected. These checklists list the most common defects; they are gradually enriched, and they constitute a history of the usual errors.

The visual inspection does not require any investment, apart from the hours spent. It allows early detection of defects or configurations that could cause difficulties during integration testing, or during the actual use of the system.

The inspection technique helps to improve the maintainability of the system via the properties of readability, simplicity and modularity.

Detailed information on the practical implementation of inspections is given in section 5.3.2.

Modelling technique

Modelling is used to represent and study one or more aspects or characteristics of the system. They are used for feasibility studies, for the definition of requirements, architectures and design. They may be static or dynamic; in the latter case they are often called *simulation*.

Dynamic models are models that represent the behaviour of the system. They are developed using computer applications, based on requirements, architecture, design, or the system itself.

From a verification viewpoint, these models are useful for verifying the consistency and completeness of functions, their inputs and outputs, command - control of the system, operational scenarios, etc. They can also serve as a reference for defining and developing functional test cases and for determining the expected results of these test cases.

Some examples of dynamic modelling:

- executable graphs
- Petri nets
- finite state-machines
- executable state-transition diagrams (state-charts, state-machines)
- executable eFFBD (extended Functional Flow Block Diagrams)
- decision-tables including *what-if* simulation

Mock-up technique

Mock-ups (or demonstrators) are physical constructions representative of a characteristic, or set of characteristics of the system. In general, mock-ups are used in the system definition stages, to **validate** an operational concept or a technological concept, a feasibility study, technical requirements, a choice of architecture, a choice of design, etc.

- Mock-ups can be static, for example:
    - MMI (Man Machine Interface) mock-up to study the ergonomic positions of monitors, screens, LEDs and commands for system operators or users,
    - wood or polystyrene mock-up to validate volumes,
    - maintenance suitability demonstrator - see Note,
- or dynamic :
    - automatism mock-up to analyse a regulation, monitoring, surveillance chain from sensors to actuators,
    - normal operation demonstrator.

Note: A maintenance suitability demonstrator (static and/or dynamic) is used to situate an operator equipped with the necessary procedures and tools for maintenance operations and to verify that these operations are feasible within the time allowed and with the provided means (presence of traps, clearances, open spaces, etc.).

Test technique

Test is a dynamic verification and validation technique. Tests are part of the experimental justification.

From a validation viewpoint, the purpose of the tests is to determine the **conformance** of the system (product, service, organization) to its requirements (stakeholder requirements and system technical requirements). This category is often referred to as **conformance testing**. We essentially distinguish:

- **Functional testing**, which consists of determining that the system performs the expected services (functionality) as required, through its functions and input-output flows. The system is seen as a black box; i.e. by applying defined inputs, the system produces the expected (specified) outputs.
- **Effectiveness testing**, which consists of determining that the effectiveness associated with the functions belong to the space delimited by minimum and maximum values as specified, and more broadly, that the objectives of the system associated with its mission are achieved.
- **Environmental testing** relating to the environment to which the system and its components are subject: mechanical environment (shocks, accelerations, vibrations, etc.), climate (temperature, pressure, sand and dust, etc.), electromagnetic (EM field, static electricity, lightning, etc.), biological, etc.

Chapter 5.1. Verification and validation of the system: Concepts and principles

From a verification viewpoint, the purpose of the tests is to reveal **defects** in the realized system. This category is often referred to as **defect search testing**. The comparison reference includes not only the system technical requirements, but also the description of architectural characteristics, design and implementation properties. The system is considered transparent for the persons in charge of the tests (glass box). We distinguish:

- Structural testing, which consists of determining the absence of defect, or searching potential defects, in the structure of the system; that is to say in the interfaces between components of the system, the interfaces between the system and the components of its context of use, in the physical components of the system (system elements).

- Capability testing, which consists of determining the absence of defect, or searching potential defects with respect to the characteristics or properties conferred to the system during its architecture and design definition. For example, the treatment of failures with regard to dependability, ergonomics or human factors with respect to usability, dimensions, compatibility between components with respect to emerging properties.

Whether it is system-product, system-service or system-organization, the exhaustiveness of the tests in relation to all its possible inputs and their combination is inaccessible. Therefore, the issue of detecting and eliminating defects necessarily leads to the use of several types of verification and validation techniques.

Notions related to testing techniques and detailed information on the practical implementation of testing are given in section 5.3.4.

Review technique

Review is a **validation** technique, which consists of gathering expert opinions on the results of one or more activities in relation to their inputs and making **recommendations**.

Reviews are similar to inspections. There are different types of reviews in the industry, for example:

- Management review: consists of analysing and evaluating the company's activities in relation to its objectives, strategy, opportunities, business models, quality policy, finances, etc.

- Contract review: consists of analysing each clause of the contract, deducing the technical, managerial and economic impacts; it is held before the signature of the contract, between client and supplier, and can give rise to negotiations.

- Project review: consists of analysing and evaluating the project's status on the management elements: calendars, deadlines, budgets, project risks, tasks execution, resource management, procurement management, quality management, etc.

- Technical review: consists of analysing the technical elements relating to the system-of-interest (documents, models, study reports, etc.) and to evaluate the maturity of the system engineering. It also makes it possible to check the conformance of the technical content to the inputs, the standards, the application of the methods, etc.

- End-of-activity review: this is a technical review held during the development of the system at the end of an activity or a group of activities. End-of-activity reviews are generally associated with predetermined milestones in the development project. An end-of-activity review is a set of verifications in order to validate the results of this activity in relation to its inputs.

The end-of-activity review technique and its implementation are detailed in section 5.3.3.

Chapter 5.1. Verification and validation of the system: Concepts and principles

5.1.4 Overview of verifications and validations

The complexity of systems is directly related to the intrinsic complexity of the problems they solve. A complex system consisting of several tens of thousands of interacting components, receiving more or less random combinations of external stimuli in a more or less hostile environment, will always show residual defects during its utilization - see the definition of *defect* and the origin of failures in section 3.5.3. To pretend to *zero residual defect* is illusory. That is to say that the elimination of all the defects is almost impossible; it is not technically conceivable and temporally imaginable, for they must be sought or detected. Most of the time, it is not economically affordable because of the necessary means.

Consequently, a complex system must live with residual defects, maintaining a satisfactory quality of service. To do this, the impact of defects solicitation (errors, failures) must be foreseen and the system must tolerate these impacts; i.e., to be robust to errors and failures, insofar as their consequences are unacceptable. This approach is the subject of Volume 5 of this series of books.

Nevertheless, verification and validation actions aim to minimize the number of non-conformances and residual defects. Validation and verification concerns have gradually led to the use of development cycles models - see Chapter 6 of Volume 1 of this series, System Notion and Engineering of Systems [Faisandier1 2015]. Their principle is to check the results of the execution of an activity or a task before going on to the next activity or task. This is obvious when realizing material objects; it is less so for abstract objects like systems, but the result is no lesser!

Therefore, the four questions that come to mind are:

1. **What must be verified and validated?**

    - The answer to this question is the subject of the verification and validation strategy, which is described in section 5.1.5.

2. **How to do it?**

    - The answer to this question was raised earlier in section 5.1.3; it is supplemented by the process approach described in Chapter 5.2.

3. **When to verify and validate?**

    - The answer to this question is given, in its principle, in the present chapter, hereafter.

4. **Who is in charge of verifying and validating?**

    - The answer to this question is the subject of section 5.3.5, which deals with the organization of verification and validation activities, as well as responsibilities.

The technical know-how of development lies in the faults avoidance, which consists of avoiding the introduction of defects and eliminating the defects introduced during the development of the system (i.e. during its definition and implementation). To this end, the principle of development cycle models is to decompose the development, on the one hand, into technical processes, themselves broken down into activities and tasks, and on the other hand, to decompose the corresponding project into phases (temporal periods) separated by milestones, while performing verification and validation actions during and at the completion of each phase, process, activity, task using appropriate techniques (listed in section 5.1.3). Figure 38 illustrates this approach.

Chapter 5.1. Verification and validation of the system: Concepts and principles

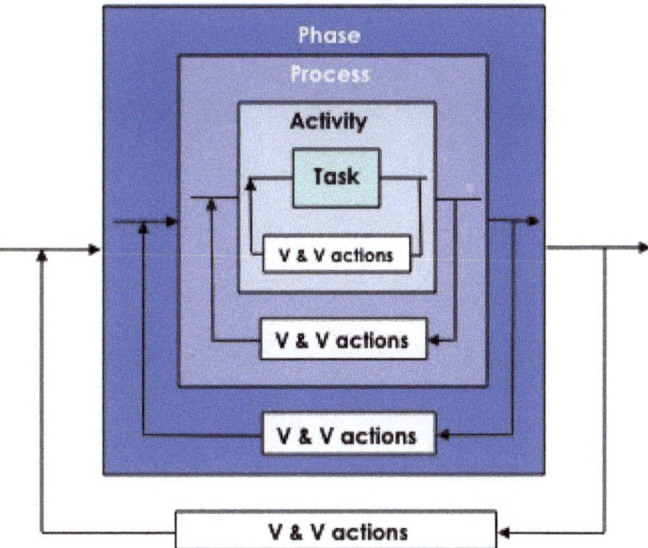

Figure 38 - When to verify and validate?

Verification and validation of a Component

Before describing the different categories of verifications and validations, and their respective places in the development of a system, we describe the case of a Component (System element). This is simpler for understanding, and it helps to make the link with practices familiar to most readers.

We consider that a component is composed of several parts. We use a V-shape structure to place the development activities. The **definition** activities are placed on the left branch of the V, the **integration** activities are on the right branch of the V. The **implementation** activities are placed at the bottom of the V; these relate to the parts, made in one or more technologies. The implementation activities consist either of manufacturing the parts, or of reusing previously developed parts, or of purchasing existing parts. See Figure 39.

The V-shape structure will remind some readers of the development cycle model known as V model. But here, this is not the purpose. The arrangement of the activities in the V-shape is used to recall the symmetry of this shape.

> The symmetry of the V-shape means that the integration and final validation activities (downstream), which concern implemented elements, are only the **consequence**, the result, of the activities carried out upstream during the definition of the Component!

The same applies to a complete system. As for the actual sequencing of activities during a project, it is done globally in an iterative way - see Chapter 3.2. The reader will find practical information on the execution of activities in the context of a development project in Chapter 6 of Volume 1 of this series of books, System Notion and Engineering of Systems [Faisandier1 2015].

Chapter 5.1. Verification and validation of the system: Concepts and principles

Internal verifications for each activity

The internal verification actions are symbolized by orange dotted circular arrows in Figure 39.

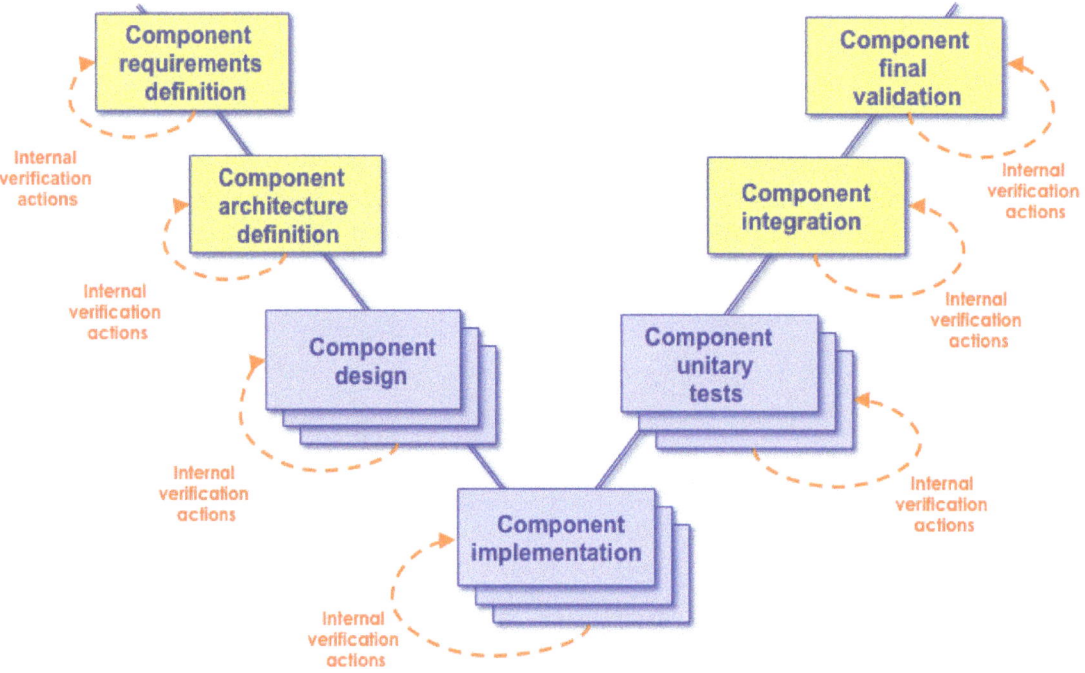

Figure 39 - Internal verifications for activities

Internal verification actions are **static** verifications to determine that planned activities are executed correctly (respect of processes, methods, standards, templates, etc.), and that the results or objects obtained are free of defects or lack thereof.

These are verification actions, not validation actions; examples:

- for a hardware component during implementation: verification of compliance with manufacturing procedures

- for a software component during coding: the compiler checks that the source code conforms with the language used; but the code could have no use (no value) - ditto for coding rules

- when defining requirements or architecture: verification in the documentation produced of the correct use of methods, tools, rules, standards

- during integration and final validation tests: verification of the existence of test cases, test execution procedures, and test results; verification of the implementation of the methodological guidelines.

Chapter 5.1. Verification and validation of the system: Concepts and principles

End-of-activity validations

The validation actions are symbolized by green dotted circular arrows in Figure 40. The red butterfly valves are decision gates placed at corresponding project milestones.

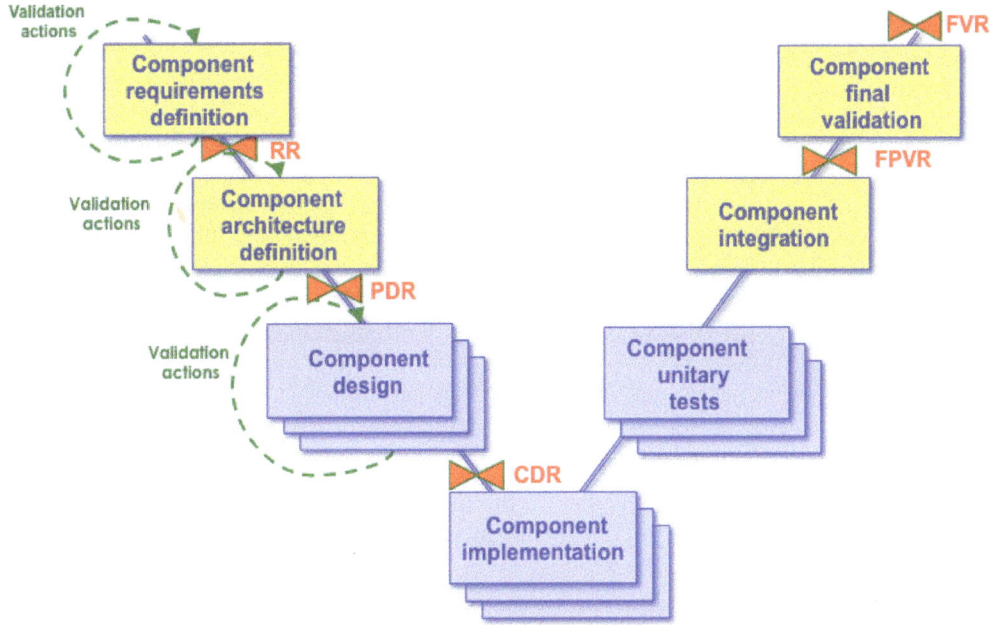

Figure 40 - End-of-activity validations

Validation actions at the end of an activity (or at the end of a phase) are **static** actions. Their purpose is to determine whether the results of the activity (of the phase) correspond with the inputs of the same activity (same phase) or not; they relate to documentation. The inputs of the activity (of the phase) are the results, the outputs, of the previous activity (of the previous phase).

These validation actions are performed using the review technique - see sections 5.1.3 and 5.3.3. In the present case, these reviews are often named **joint reviews**. They gather, during project milestones, the customers of the activity (phase) and the suppliers of the activity (phase) that ends.

The purpose of these reviews is to validate the results of the activity (of the phase); a value judgment is made on the elements analysed to decide whether or not development activities can continue, with or without corrective action. To be relevant, this value judgment requires two conditions: the **expertise** of the technical field and the subject analysed, and the **exteriority** to the project, because one cannot be judge and part at the same time.

The development project team provides the external experts with the justifications for the work carried out in order to assess the validity of the documents produced. For example:

- At the **RR milestone - Requirements Review**, carried out when the Component requirements are considered sufficiently mature: the requirements (output of the activity) are analysed to check, on the one hand if they correspond to the real need or perceived need (inputs of the activity), and on the other hand if their respect by the implemented Component will be verifiable; in other words, whether it is possible to define a validation method and technically feasible means, economically compatible with project budgets.

- At the **PDR milestone - Preliminary Definition Review**, carried out when the architecture and design descriptions are considered sufficiently mature: the architecture of the Component and the design properties (outputs) are analysed to check, on the one hand if these elements satisfy the Component requirements (inputs), and on the other hand if the architectural and design choices are properly justified.

- At the **CDR milestone - Critical Definition Review**, carried out when the design of parts of the Component is considered sufficiently mature: the detailed properties and design plans (outputs) of each part are analysed to check, on the one hand if these elements conform with the architectural characteristics and design properties (inputs) of the parent Component, and on the other hand if the parts are implementable and can be integrated together, especially with regard to their interfacing.

- At **FVPR milestone - Final Validation Preparation Review**, carried out during or at the end of the integration of the parts to form the Component, but before any final validation activity: each implemented part and its validation documentation are analysed, to check if they are available and validated (conform with their definition documentation) and the means required for final validation are also available and validated.

- At the **FVR milestone - Final Validation Review**, carried out at the end of the final validation activity: all supplies relating to the Component are analysed to check if they are correct, complete and validated, and if these are available for delivery or for transfer of ownership to its user or customer.

Note: The names of the reviews and corresponding milestones used above are not standardized; these are just examples - see Chapter 6 of Volume 1 of this series of books, System Notion and Engineering of Systems [Faisandier1 2015] for more information about development cycle models, phases, milestones and decision gates.

Dynamic experimental verifications and validations

Dynamic experimental verification and validation actions, i.e. **tests**, are carried out during the activities on the right-hand branch of the V. They are symbolized by dotted arrows in Figure 41: dotted violet arrows for verification actions, dotted orange arrows for validation actions. These actions consist of checking that the implemented Component conforms, or not, its definition references. The definition references of the Component are elaborated during the execution of the definition activities on the left-hand branch of the V; this explains why the violet arrows start from the right to go to the left.

The black continuous lines symbolize the passage of the tests definitions and plans from left to right; because the tests are defined and prepared from the Component's definition references, and these activities are carried out in parallel with the definition activities of the Component.

Dynamic experimental verification and validation actions (tests), carried out by the integration and final validation activities, aim to check the adequacy, or not, of the Component with respect to its definition references:

- verification and validation of each part of the Component with respect to its design,
- verification and validation of the Component with respect to its architecture and design properties (functional, physical and interfaces aspect),
- verification and final validation of the Component with respect to its requirements (functional, effectiveness, interface, etc., requirements).

Note: The procedures for implementing the tests depend on the means necessary for the execution of the tests. They are elaborated from the tests definition and plan, in parallel with the development activities, but before the tests are carried out.

Chapter 5.1. Verification and validation of the system: Concepts and principles

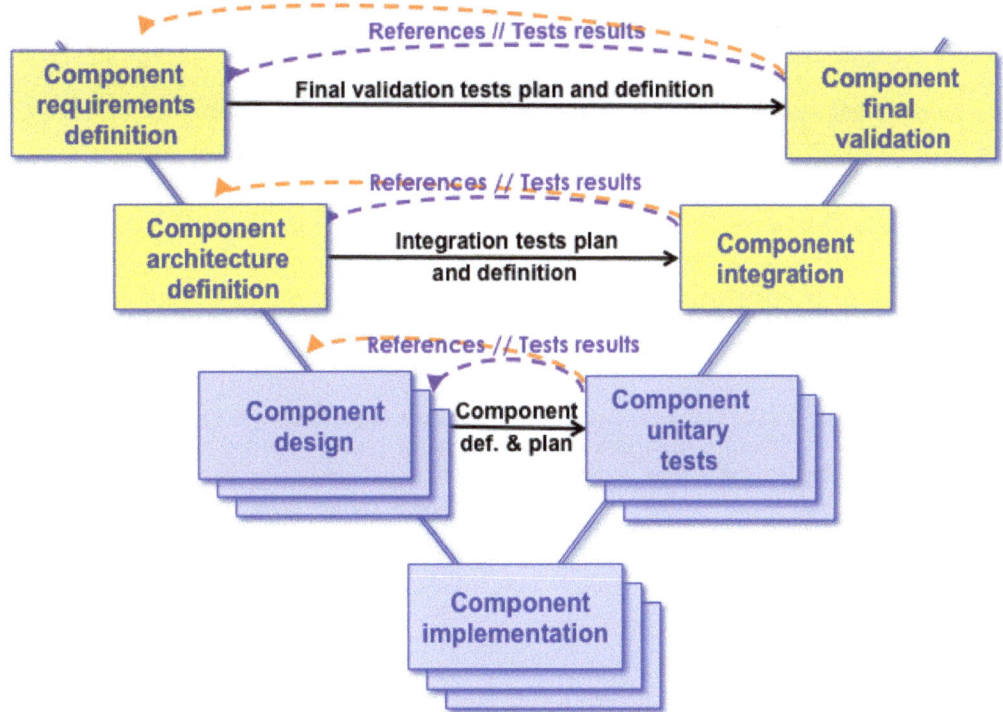

Figure 41 - Dynamic experimental verifications and validations

Verifications and validations synthesis

To sum up, the different categories of verification and validation actions concerning a Component are as follows:

- Internal **static verification** carried out during each task of the activity to check the consistency of elements produced during this activity; these verifications deal with the form and the application of related professional standards (methods, techniques, tools, rules, etc.) - see Figure 39.

- End-of-activity **static validations**, named *end-of-activity reviews*, to check the consistency of elements produced during this activity with respect to outputs of the previous activity - see Figure 40.

- **Dynamic verifications** carried out on the Component during the integration and final validation activities; these verifications deal with the adequacy, of parts then the Component, to corresponding definition references - dotted violet arrows in Figure 41.

- **Dynamic validations** carried out on the Component during the integration and final validation activities; these validations deal with the conformance, of parts then the Component, to corresponding inputs of definition activities - dotted orange arrows in Figure 41.

Note: A fifth category of verification could be added: quality verification. It checks compliance of activities outputs with verification and validation plans and procedures; for example, review procedures, tests procedures.

Chapter 5.1. Verification and validation of the system: Concepts and principles

Verification and validation of a System

It is impossible to perform a single, global verification or validation action directly on a fully integrated complex system. The errors and defects observed during this global observation can be numerous; identifying the causes of non-conformance is almost impossible.

It is partly for this reason that the system-of-interest is decomposed, when it is defined, in layers of (sub) systems and components (system elements), globally in top-down order, applying the notion of system-block - see section 3.3.1. The upper layer system requirements are allocated to lower layer (sub) systems, during architecture and design definition activities, with traceability links to ensure the gradual satisfaction of system requirements, and to ensure consistency of architectural characteristics and design properties of the set. Verification and validation actions are then applied to the definition of each (sub) system, as described above for a Component (a Component can be seen as a (sub) system).

Conversely, the implemented system (product, service, organization) is obtained in a bottom-up way during the integration, by **composition** of the (sub) system starting from the most elementary Components up to the global system. Verification and validation actions are applied to each Component or (sub) system so that it is verified and validated, possibly corrected, before being integrated into the parent (sub) system of the upper layer - see illustration Figure 42.

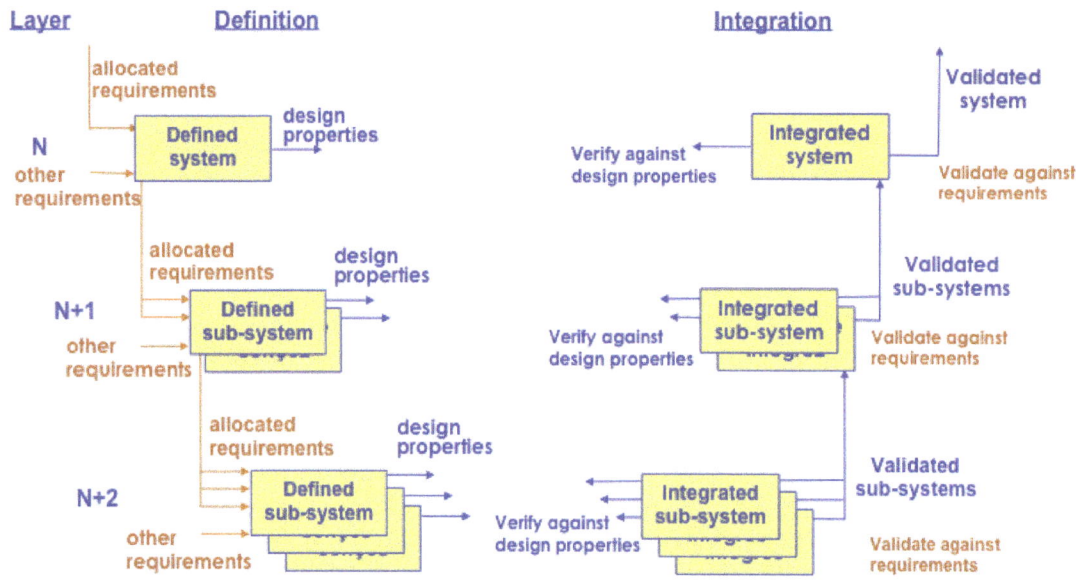

Figure 42 - Verification and validation of the system layer by layer

Note: In Figure 42, the term *allocated requirements* means requirements applicable to the (sub) system; the term *other requirements* includes the following set: stakeholder requirements of the layer, applicable normative requirements, derived (induced) requirements.

If necessary, the (sub) systems, the elementary components are partially integrated into subsets or aggregates (see section 6.1.1) in order to limit the number of properties to be verified in a single step. For each (sub) system of each layer, it is necessary to perform a set of final validation actions to verify that the characteristics of the lower layer are not damaged.

In addition, a conformed verification result, obtained in the environment of a layer, can turn into not conformed in the environment of a higher layer. As long as the system is not fully integrated and/or does not function in the actual operating environment, no results must be considered as definitive.

5.1.5 Verification and validation strategy

Synergy of strategies

The beginning of section 5.1.4 explains that a complex system potentially contains defects. These defects are introduced during development because of the nature of the human activity of creation and realization - see section 3.5.2. In order to reduce their number, which is obviously unknown, and to increase confidence in the use of the system, we have seen that it is necessary to carry out verification and validation actions throughout development. Not only during the integration of the physical system and its components, but also during the definition of requirements, architecture, design and during implementation. Given the number of elements created and the number of relationships between these elements, it is easy to understand that the number of verification and validation actions to be defined is even greater. The question then arises as follows:

- Is it possible to verify and validate **everything**, given the limitations imposed by any project? Namely a delivery deadline, defined budgets, finite resources.

The answer is probably negative in the case of a complex system development project; the next question is:

- Which verification and validation actions must be defined and executed at a minimum, to minimize the risks incurred if certain actions are ignored?

The answer to these questions is the subject of a **verification and validation strategy** adapted to the system to be developed.

We will see later that there are two distinct processes, a verification process and a validation process. There are therefore two strategies, one verification strategy and one validation strategy. We have previously seen, in section 5.1.2, differences between verification action and validation action. In reality, these actions are very similar, because they use the same techniques or methods (described in section 5.1.3); any validation action is preceded by verification actions; and finally, if both actions contain a conformance aspect, only the validation action decides whether the conformance gap is acceptable or not.

From a practical viewpoint, there is no need to distinguish between verification action and validation action, because the essential difference lies in the reference of comparison between the expected result and the obtained result: the verification of a system uses the architectural characteristics and design properties of the system as references, whereas validation uses as references the allocated requirements of the higher layer - see Figure 42. Just as architectural characteristics and design properties are linked to system requirements, verification actions are linked to validation actions.

Finally, by anticipating the next chapter, the activities of the verification process and the activities of the validation process contribute to the same objective, i.e. the quality of the implemented object. The actors are different, their goal is similar, but their approach could follow different paths:

- The client wants that the implemented object is the one envisaged, favouring global validation actions.
- The supplier seeks to clear defects introduced during development by focusing on more detailed verification actions.

These last considerations show that it is more efficient and optimal to develop a single strategy, named verification-validation strategy, implementing synergies using identical techniques, methods and means.

Elaboration of the verification - validation strategy

The preamble of this strategy is to identify or determine:

- the scope of verifications and validations; i.e., to determine all that **SHOULD** be verified and validated, and to give objectives in qualitative and quantitative terms - this point is detailed below in this section;

- limiting or imposed constraints; i.e., identify the constraints that will limit the number of verification and validation actions - this point is detailed following the precedent in this section.

In practical terms, developing the verification-validation strategy consists of answering the following questions; the same questions are also asked for each verification and validation action:

- **What**? and **How many**? This makes it possible to define the scope.
    - What elements of engineering, implementation and integration should be verified and validated? With what coverage rate?
    - Which verification and validation actions should be performed on these elements?

- **How**?
    - What methods or techniques should be used?
    - What tools?

- **When**?
    - When should these actions be carried out (project phase)? Planning
    - In which order? Scheduling
    - At what time? Calendar
    - What are the criteria for deciding whether to stop verifications and validations?

- **Where**?
    - Where do these actions take place (laboratory, plant, real site, etc.)?
    - With what means (tests bench, measurements bench, etc.)?

- **Who**?
    - What are the responsibilities and the organization?

The elaboration of the verification-validation strategy is one of the activities of the processes described in sections 5.2.1.3 and 5.2.2.3.

Chapter 5.1. Verification and validation of the system: Concepts and principles

Scope of verification - validation

The perimeter of the verification-validation defines the elements that must be subjected to verification and validation actions. These elements are instances of almost all entities in the development meta-data model; for example: stakeholder requirement, system requirement, design property, function, flow, component, physical interface, scenario, physical architecture, etc.

These elements are usually grouped into documents and models; for example

- Contract, customer request, specification, operator handbook:
    - needs or expectations, stakeholder requirements, system requirements
    - quality, commercial, management clauses, for example: joint review, final validation, guarantee, acceptance, utilization of operational scenarios including customer participation, participation of users to verification - validation actions, expressed quality characteristics
    - certification
- Imposed standards and regulation
- Description of the architectural solution:
    - functions, flows
    - functional, behavioural, temporal architecture
    - components, physical interfaces
    - structural physical architecture, outline drawing, volumes, dimensions
    - design properties
- State of the art:
    - standards, professional references
- Enterprise policies:
    - enterprise processes
    - standardisation policy
    - reuse policy

The list of elements to be verified and validated can be very long; it is almost impossible to verify and validate everything. Insofar as correct engineering processes (such as those mentioned in the volumes of this series, certain standards or handbooks) are actually used in a development project, qualitative and quantitative objectives can be set.

For **qualitative objectives**, it can be assumed that each system requirement (coming and traceable from stakeholder requirements) must be *valid-able*. A requirement is *valid-able* if there is a suitable technique or method that is economically affordable, to check that the system (product, service, organization) respects this requirement.

But for a requirement to be *valid-able*, its expression must first be verified and validated. Then, it will be possible to verify and validate that the system respects the functional, effectiveness, interface, operational conditions (maintainability, usability, interoperability, integrity, safety, etc.) requirements, the physical, implementation, maintenance, production, disposal constraints, etc.

During the system definition, system requirements are gradually transformed into architectural characteristics, then into global design properties for the system or allocated to (sub) systems or components. Each characteristic and each property can be verified; more precisely, their respect by the system and its components can be verified. They are elements of the logical architecture (function, flow, trigger, mode, transition, performance, etc.) and of the physical architecture (component, physical interface, dimensions, reuse rate, coupling rate, availability, reliability, access control, etc.).

For **quantitative objectives**, the notions of coverage and coverage rate are used:

- **Coverage** of element E is obtained when this element has been verified or validated at least once with an appropriate technique or method. For example, *requirement R has been covered*, means: the respect of requirement R has been checked at least once with an appropriate technique. Another example: *the path coverage of model M is effective*, means that all paths of M have been exercised and verified at least once.

- The **coverage rate** of the E elements is equal to the number of E elements verified (or to be verified) / the total number of E elements. For example, the functional coverage rate is 100%, means that each functional requirement has been verified or validated at least once.

Coverage is obtained through static, dynamic verifications or by tests.

Coverage of performance or effectiveness requirements is difficult to obtain for at least two reasons: it is not always possible to precisely define a measure of effectiveness, because of its subjective and/or negotiable aspect; the means required to check them are sometimes complex and very expensive to implement. This is why we are talking more about effectiveness evaluation than obtaining coverage. The method consists of:

- performing theoretical studies to define the expected measure of effectiveness and the verification framework (worst case, checkpoints, etc.),

- verifying the actual measures of performance, onto the implemented system or its components,

- justifying the gap if any.

> What coverage and coverage rates are reasonable to achieve during verifications and validations?
>
> A general recommendation is to achieve 100% coverage of functions, interfaces and dynamic behaviour, limited to specified operational scenarios.
>
> In addition, use structural coverage (of any structural elements of the architectures, seen in glass box) to check the robustness and to detect the functional verifications forgotten.

Structural coverage is often used as a criterion to stop verifications, regardless of the physical system decomposition layer. It is advisable not to focus on the structural coverage: it should only come in addition to the functional coverage (components seen as black boxes).

The effort to implement structural coverage is very important; it is why they are often limited to dependable applications, applications requiring robustness, or critical and complex components. They require specific means for the implemented technologies.

The recommendation set out here is reasonable in that the engineering has followed the state of the art. Indeed, through the upstream traceability matrices, system requirements / stakeholder requirements, it is ensured that the stakeholder requirements have been met (verified). Through the downstream traceability matrices, it is ensured that the characteristics and properties given to the system correspond with the system requirements (but not necessarily all, because architects and designers are obliged to add functions, implementation constraints, devices in case of errors or failures of the components, information that does not necessarily have a place in the requirements).

Limiting constraints

The constraints that limit the execution of verification - validation actions come from the following parameters:

Time

What is the time available to carry out the verification - validation actions against the project deadlines or the deadline for delivery? For a complex system, it would take more time to carry out all the possible verification - validation actions than to define, conceive and implement the system itself. What is the time needed to achieve all the means associated with these actions? What is the availability of resources and tests means? Etc.

Cost

What is the budget assigned to the execution of the verification - validation actions? What is the budget assigned to the associated means?

It is possible to reduce the overall cost of verification and validation by focusing on actions that can be carried out most upstream of the project; for example, during the definition of the system because the verification - validation actions require few means compared to those carried out downstream like the tests. Some architectural characteristics, such as modularity, reuse, standardization of interfaces, etc., contribute to the reduction of the number of verification - validation actions, and therefore costs. It is also possible to increase the effectiveness of the tests.

Industrial organization

Is it possible, and how can we distribute the execution of verification - validation actions among the organizational units concerned by the project: customer representative, user representatives, architects, developers, subcontractors, and suppliers? What is the complexity of the relationship between these units? Is it possible to share resources and means with other projects?

Criticality of the system

The criticality of a system depends on its mission, on the safety aspects of this mission, on the selected technologies and/or on the maturity of the participants. Dependable systems require more complete coverage.

Regulation

What is the possibility, the impossibility or the prohibition of destructive tests, dangerous for the environment and/or for the personnel in charge of their execution? Are the means certified, etc.?

Means

The means associated with verifications and validations induce the strongest constraints for feasible verification - validation actions. Verification - validation actions use many and different types of means:

- Tools to assist with theoretical verifications, experimental verifications (without testing), definition of test cases
- Instrumentation tools or means for capturing data, signals and measurements
- Tools or means for development and management of verification - validation procedures
- Tools or means for analysing and archiving results
- Tools or means for developing test data sets (input / output flow generators, simulators, etc.)
- Tools or means for executing tests (test benches, infrastructures, runways and test grounds)

Chapter 5.1. Verification and validation of the system: Concepts and principles

It is therefore sought to optimize these means by using the development means dedicated to the target system, and/or separate means (i.e. means common to several systems of the same nature or not, several projects), and/or available operational means (e.g., elements of the context, components of enabling systems). This optimization requires an analysis of the independence of these means, and decision criteria such as cost, deadlines, feasibility of means, sustainability and availability.

For example:

- Independence between the development means of the target system and the means for executing the tests in relation to operational means; the selection criterion is the availability of operational means; in the case of serial production following the development, the means for verification - validation will probably be separate means.

- Independence between the development means of the target system and means for executing the tests; the selection criteria are the cost of the separate means, the time required to obtain these means, the compatibility or conformance of the environments for use.

Often, the means for testing are technologically more complicated to achieve than the target system itself. However, if it is not possible to develop the test means, it will also be impossible to verify or validate the characteristic or the design property or the corresponding requirements.

> The identified constraints are used later to justify not retaining certain verification - validation actions. Because, most of the time, it is impossible to verify - validate all the properties and their combinations exhaustively within compatible times and budgets of the project's resources.

Selection of verification - validation actions

The verification - validation strategy is deduced from the comparison between the objectives (what we **SHOULD** verify and validate) and the limiting constraints; i.e. the definition of the verification - validation actions that **CAN** be reasonably undertaken, at what time and by what means.

The selection of verification - validation actions depends on the type of system, the objectives of the project, but above all the acceptable risks; i.e., risks that are accepted if a verification or validation is not performed. The verification - validation strategy includes optimization by selecting the best combinations of verification - validation actions to be carried out with regard to costs, deadlines, efficiency and risks.

Here are some simple rules to make a first sort:

- Eliminate, replace and justify verification - validation actions that do not satisfy the regulatory, organizational, cost, time and resources constraints
- Eliminate redundant verification - validation actions
- Every validation action must be preceded by verification actions
- Every test during the integration phase must be preceded by corresponding verification actions during the definition phase

When establishing the strategy, for the selection and execution order of verification - validation actions, it is recommended to proceed as follows:

- Begin with static verifications as early as possible, whatever the development activity and the type of system, at any level of the decomposition of the system into Components
- Continue with functional tests (black box) at the level of Components of the system, then complete with:

Chapter 5.1. Verification and validation of the system: Concepts and principles

- ♦ grey box type tests to improve functional and structural coverage
- ♦ glass box type tests for particular aspects, for example robustness
- Continue with integration tests when integrating the Components of the system, to achieve coverage of functions and interfaces
- Complete with structural type tests to achieve other architecture, design and implementation arrangements
- End up with system validation tests and integration into the higher system level

Note: As explained in section 5.3.4.1, tests can be classified in three categories:
- ♦ Functional tests, **black box** type, relating to a function (actions, transformations, executed treatments) including in non-nominal cases
- ♦ Structural tests, **glass box** type, relating to the structure of the Component or the system for non-specified and/or unplanned behaviours during the architecture definition, but incorporated during design definition or during implementation
- ♦ Structural tests, **grey box** type, relating to the structure of the Component or the system for non-specified behaviours (non-nominal), but planned during the architecture definition (and thus expressed in engineering documents)

Planning

Once the list of verification - validation actions, that will actually be carried out, is completed, the most relevant method or technique of verification - validation (e.g. demonstration, and/or inspection, and/or testing) must be determined for each action; then the means necessary for carrying out the verification - validation actions according to the selected verification - validation method or technique: physical means (test bench, mock-up, measurement device, etc.), qualified personnel, consumables, various facilities, energy, location or infrastructure (e.g. wind tunnel, basin, cold environment, etc.).

When to perform verification - validation actions during development? The overview of verification and validation actions in section 5.1.4, and in particular Figure 39, Figure 40, Figure 41 and Figure 42, provide some answers to this question; because each development activity must be associated with one or more verification - validation actions. More precisely, for a given development activity, it is necessary to define:

- the verification - validation actions to be carried out
- on which objects or entities
- the trigger events
- the means required for their execution
- the criteria for stopping verification - validation actions
- the sequence of verification - validation actions

Planning for verifications and validations must be consistent and integrated with the planning of development tasks; the activities of the verification and validation processes must be mentioned macroscopically in the system development plan or in the corresponding project management plan.

The definition of the strategy includes the scheduling and timing of the execution of verification - validation actions; these actions and their implementation are detailed in a verification - validation plan - see Annex chapter 8.2. The following recommendations made it possible to improve the effectiveness of the verification - validation strategy on real projects:

- Plan verification - validation actions on the upstream activities of the project, knowing that a modification on the engineering (documents, or plans) is much cheaper than a modification on a physical prototype in integration.

- Try to group together the maximum number of verification - validation actions over time and on the same configuration (e.g. tests on a test bench).

- Verification - validation actions must validate the results of a phase (or development activity) before proceeding to the next phase or activity. They must appear in the project calendar, with their trigger event, not forgetting the tasks upstream of the execution for the definition and the launching or the preparation of these actions, and the tasks downstream of the execution for analysing the results.

- In the planning, do not forget the "compulsory windows" in time, for example for the use of test means that are only available for a given period.

- Do not forget to plan the implementation, or the purchase of the means necessary to carry out the verification - validation actions. As such, for complex systems, it may be useful to define one or more enabling verification - validation systems, which can be the subject of a resources / means development project; a separate project but coordinated with the project associated with the system-of-interest.

Indicators for monitoring verification - validation actions

In the verification - validation strategy, it is also necessary to define the indicators that will make it possible to determine, at any moment of development, what has been verified, and especially **WHAT HAS NOT BEEN VERIFIED AND WHY**. Because it is important to know why verification - validation actions have not been carried out, and to assess the risks involved.

Example of indicators:

- Verification - validation progress indicators = ratio between the number of verification - validation actions performed and the number of verification - validation actions selected; this calculation can be done per phase, per development activity or globally on the project.

- Conformance rate indicator relating to the object submitted to verification - validation.

- Verifications performance measures: detection of gaps, non-conformances, magnitude of these gaps? Are the means used proportionate to the defects detected?

- Functional and structural coverage rates measures.

Stop of verifications

The purpose of verification and validation is to establish the conformance of a product, service or organization and to search potential defects.

But carrying out unlimited verifications leads to drift risks for costs and deadlines. Indeed, back and forth between verifications and modifications follow normally until the system is reasonably compliant and free of defects.

The absolute zero defect is a utopia, especially if the system presents a huge combination of potential behaviours, predictable or not!

When establishing the verification - validation strategy, and more precisely when defining the verification actions, it is necessary to fix:

- limits for cost, deadlines, or the maximum number of back and forth,
- criteria to stop based on indicators or measures, such as percentage of success, number of defects detected, coverage rate obtained - see Note, etc.

These limits are established on the basis of feedback from similar projects, taking care to allocate margins in the development or the use of innovative technologies or components.

Note: As a reminder, it is important to have, at any time, a precise idea of the verification and validation actions that have not been carried out and for what reasons, in order to assess the risks incurred. The current coverage rate [= Number of items verified and validated / Total number of items] can be compared to the objective coverage rate [= Number of items to be verified and validated / Total number of items].

Chapter 5.2.1. Verification of the system: Process approach

5.2 Process approach

As a reminder, as mentioned in the section "Verification and validation approach" - see 3.3.4, the verification and validation processes are support processes that are used by the definition and integration of the system. These are generic processes that must be instantiated according to the objects or elements to be verified and validated. In the present case, the objects to be verified and validated are the system itself, its engineering data and the integration of its components to form the expected product, service or organization.

5.2.1 Verification process

This section describes:

- The location of the process in the development cycle - section 5.2.1.1
- The purpose, inputs and outputs of the process - section 5.2.1.2
- The activities of the process - section 5.2.1.3
- The ontology elements used - section 5.2.1.4
- The main artefacts produced by the process - section 5.2.1.5

5.2.1.1 LOCATION OF THE VERIFICATION PROCESS

Figure 43 shows the location of the verification process among the processes of definition, implementation and integration of the system. In the left part of the figure, the verification process covers definition elements, entities and artefacts. In the right part, it focuses on physical elements, entities and integration artefacts.

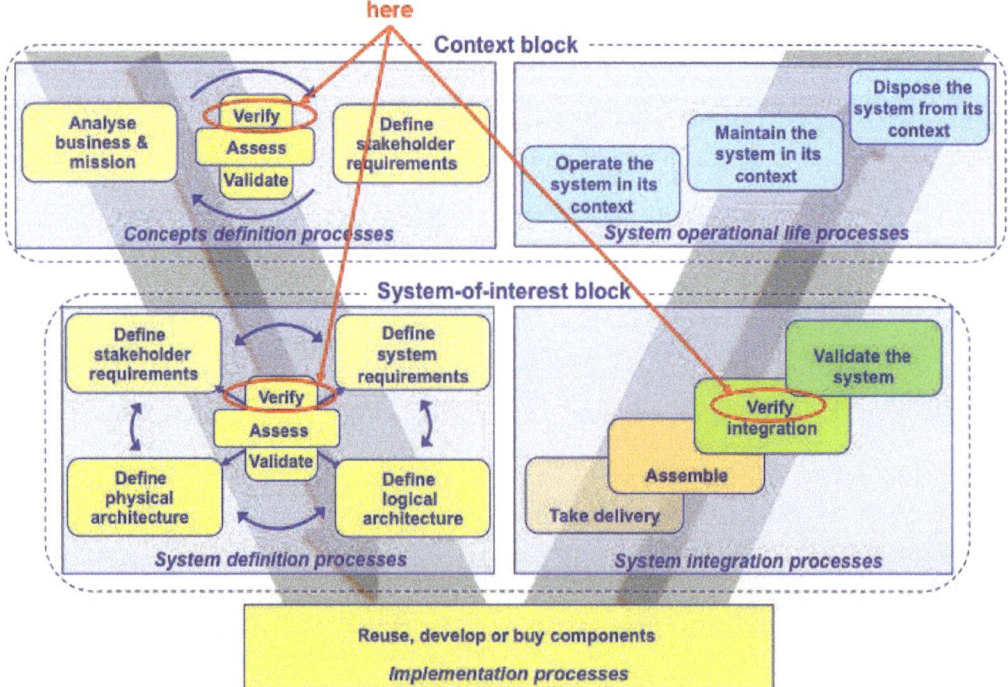

Figure 43 - Location of the verification process in the development

5.2.1.2 PURPOSE, INPUTS AND OUTPUTS OF THE PROCESS

Synthetic and generic expression

The purpose of the verification process is to determine that an implemented object* conforms with its characteristics or design properties.

The aim of the verification is to identify the defects introduced at the time of any transformation from input to output

- see 3.2 Notion of process.

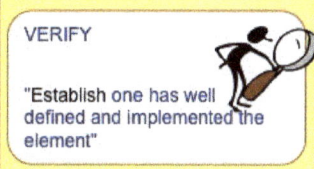

VERIFY

"Establish one has well defined and implemented the element"

Discussion

* : By "implemented object", understand any engineering entity such as implemented component (product, service, organization), procedure, document, stakeholder requirement, system requirement, logical architecture, physical architecture, design property, test procedure, etc.

The verification process is instantiated as many times as necessary during the development of the system and for each implemented object.

The verification process is used to determine that the transformation used to obtain each object was made with the selected and approved methods, techniques, standards and transformation rules. This can be achieved by checking that the implemented properties of each object match the expected properties.

Any verification is based on tangible evidence; that is to say on information whose veracity can be demonstrated, based on facts obtained using techniques, such as inspection, test, analysis, demonstration, simulation, etc. - see section 5.1.3.

To verify a system (product, service, organization) is to determine that its real characteristics or properties conform with its expected architectural characteristics and design properties. It is in this sense that the following descriptions are given, i.e., the verification process is instantiated for the definition and integration of the concerned system.

Note: The standard ISO/IEC/IEEE 15288, System life cycle processes [ISO 15288], provides the following purpose of the system verification process; the expression given in this book is therefore broader.

The purpose of the Verification process is to provide objective evidence that a system or system element fulfils its specified requirements and characteristics.

The Verification process identifies the anomalies (errors, defects, or faults) in any information item (e.g., system requirements or architecture description), implemented system elements, or life cycle processes using appropriate methods, techniques, standards or rules. This process provides the necessary information to determine resolution of identified anomalies.

Chapter 5.2.1. Verification of the system: Process approach

Inputs

The inputs of the process are:

- the object submitted to verification,
- the **definition references** of the object submitted to verification with respect to the specific characteristics of this object,
- the **transformation references** (state of the art or method) used to obtain the object with respect to the generic characteristics of this object,
- the verification means and/or tools required to carry out the verification actions.

Notes:

To verify the system definition, each engineering data must be verified; for example: **Stakeholder Requirements**, **System Requirements**, logical architecture elements / entities (**Functions**, **Input / Output Flows**, etc.), physical architecture elements (**Component**, **Physical Interface**), **Design Properties**, and so on. The reader can do the exercise of finding the **definition** and **transformation references** of each of the engineering entities and deducing the specific and generic characteristics to be verified. However, this information is indicated in the description of activities and in a section of each system definition process, in Volume 2, Systems Opportunities and Requirements [Faisandier2 2013] and in Volume 3, Systems Architecture and Design [Faisandier3 2013], of this series of books.

To verify the integration of the system, i.e. the implemented product, service or organization, the system **definition references** essentially consist of the System Architecture Description Document and the Design Document(s) of the system and/or its Components, for the specific characteristics of the system; the **transformation references** consist of Components Manufacturing Control Files (including the interfaces).

Outputs

The outputs of the process are:

- the verification plan that includes:
  - the verification strategy, or the reference to the system verification - validation strategy
  - the definition of the selected **Verification actions**
- the **Verification procedures**
- the **Verification configurations**
- the needs and/or the requirements for the **Verification tools**
- the results of the execution of the **Verification actions**
- the reports containing the analysis of the results of verifications
- the anomaly and/or non-conformance sheets

Note: The overall result of the correct execution of the process is a verified system (product, service, organization); it is an observation (verified system) and not a transformation.

5.2.1.3 ACTIVITIES OF THE PROCESS

The activities of the verification process follow:

    A. Prepare the verification

    B. Perform the verification

    C. Analyse and record the results of the verification

    D. Control the verification process

Figure 44 presents the activities of the process, the exchanged engineering data and the main artefacts generated by the process.

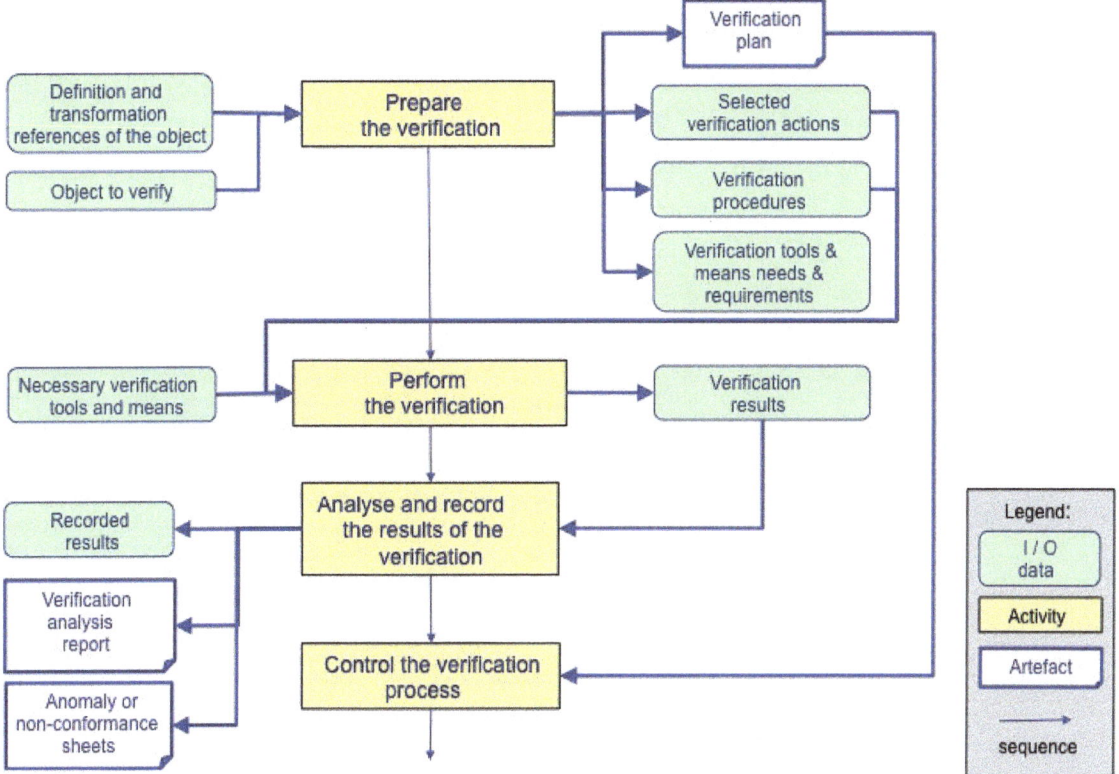

Figure 44 - Activities of the verification process

Chapter 5.2.1. Verification of the system: Process approach

Details of activities

A. Prepare the verification - This activity includes the following tasks:

1. Define the verification strategy. This task is composed of the following sub-tasks:

- Analyse the perimeter of the verification - see section 5.1.5. This analysis leads to the definition of a list of objects to be verified and of **Verification actions** to be exercised on these objects - see section 5.1.2. This analysis is normally carried out as soon as the concerned objects are defined. For example, as soon as a **System Requirement** is written or expressed, define **Verification actions** to detect whether the requirement as written is verifiable. If this is not the case, just after the execution of the Verification action, it will be necessary to reformulate or re-examine the requirement, or to break it down into several requirements; it is the same with the elements and characteristics of logical and physical architecture, and also with the **Design properties**. In this sense, the verification process exerts a significant feedback on engineering, if it is executed early and iteratively.

- Define qualitative and quantitative objectives (coverage rates - see section 5.1.5) for each object to be verified. This allows providing criteria to stop the verifications.

- Establish the list of limiting or imposed constraints - see section 5.1.5. These constraints impact the implementation of **Verification actions**, and reduce the number of **Verification actions**; example of constraints: feasibility, cost, deadline, means, safety, regulations, accessibility, measurability.

- Select the verification techniques or methods relating to the objects to be verified and the **Verification actions** taking into account limiting or imposed constraints - see section 5.1.3 (for example: demonstration, inspection, tests, simulation, etc.).

- Select the **Verification tools** or means required to perform the **Verification actions** and the associated verification techniques or methods; they are physical means (test bench, mock-up, measurement devices, etc.), qualified personnel, consumables, infrastructure, etc.

- Define the scope of the verification by comparing verification objectives, limiting constraints and the means required for verifications. This leads to the definition of a list of **selected Verification actions**, at what times it is appropriate to execute them, and by what means. Nevertheless, a risk analysis should be conducted to identify unacceptable or acceptable risks if a **Verification action** is removed from this list.

- Organize the **Verification actions** by dividing them into the various phases of development, the different levels of decomposition (system, subsystem, component), different verification techniques or methods, different **Verification tools** or means. Practically, traceability matrices (allocation matrices and justification matrices) can be used or consolidated.

- Optimize the **verification strategy** against the project deadlines and costs and the **Verification tools** or means needed, by analysing the validation strategy and the integration strategy to detect synergies. Delete redundant **Verification actions**; if possible, group **Verification actions** and **Validation actions**; postpone the execution of certain **Verification actions** from a system level to a higher or lower level to avoid developing expensive **Verification tools** or means. Promote **Verification actions** that can detect the most potential defects.

- Document the **Verification actions** in Verification Action Sheets - see section 5.2.1.5, and the verification strategy in the System Verification Plan - see section 5.2.1.5.

2. Develop the verification procedures. A **Verification procedure** generally groups several **Verification actions** on the same **Verification tool** or means. It refers to the **Verification actions** to be performed, describes the necessary **Verification configuration** (objects to be verified, initial conditions, tools, personnel skills), execution steps and stimuli, checkpoints and results to be recorded. Document in Verification Procedure Sheets - see section 5.2.1.5.

3. Identify the constraints induced by the verifications on the system-of-interest and its Components. These constraints are revealed when the verification strategy is established; they derive from the study of the feasibility or the implementation of certain selected verification actions. These constraints can give rise to System requirements, architectural arrangements (Component, Physical Interface, volume, surface, location, etc.), design layouts, and Design properties. They must be provided to the processes concerned in order to be incorporated in the corresponding references. These constraints concern, for example, performance or effectiveness characteristics to be verified, accessibility to certain components or interfaces, measures to be carried out, etc.

4. Define or specify the Verification tools or means necessary for the execution of the Verification actions and their deadlines for availability. These means can be tools, products, services, enabling systems. This task includes defining the technical requirements of these enabling elements, their interfaces with the system-of-interest, and the deadlines for delivery or availability.

5. Obtain the Verification tools or means. The means must be available at the appropriate time to be able to perform the Verification actions using the previously defined Verification procedures. The means can be obtained in different ways, by leasing, acquisition, development, reuse, subcontracting. One or more means of verification can be the object of a complete enabling system developed in the context of a project separate from that of the system-of-interest. Attention, the development of some means may be longer than that of the system-of-interest; it is a matter of coordinating developments over time.

B. Perform the verification. This activity consists of carrying out the basic operations described in the Verification Plan and in the Verification actions using the Verification procedures, with the means and the Verification tools provided, under the prescribed conditions and in the forecast calendar. More precisely:

1. Gather the objects to be verified into subassemblies as provided for in the Verification procedures.

2. Construct the Verification configurations (objects to be verified, means and tools, qualified personnel, resources or consumables required) and set up the conditions stipulated in the Verification procedures.

3. Perform the Verification procedures as prescribed, following the step-by-step procedure.

4. Collect the results obtained through the execution of the Verification procedures and the intermediate measures required by the Verification actions. Note the gaps between the defined conditions and the actual conditions of execution.

C. Analyse and record the results of verification. This activity consists of providing diagnostics and recommendations. The tasks are the following:

1. Compare obtained results against the expected results as described in the Verification actions.

   - If there is no gap, the object conforms with the expected characteristic.
   - If there is a gap, the object does not conform with the expected characteristic.

2. Record the status of the comparison in the verification Analysis Reports or Sheets provided for this purpose: conform or non-conform. In the non-conform case, an Anomaly or Non-conformance Sheet is initialized.

3. Investigate the cause(s) of the gap in the case of non-conformance. The causes can be: an error in the verified object, an error in the Verification action or in its execution, an unrealistic expected result (the references are involved), an error in the Verification procedure, in the Verification configuration, in the execution conditions, etc. If the verified object or its

references are involved, any changes must be made by the concerned processes, under no circumstances by the verification process. All other cases are the responsibility of the verification process, which must modify the concerned elements, before being able to be used again in subsequent verification sessions.

4. Establish the Verification Analysis Reports and any Anomaly or Non-conformance Sheets; record the information in the project databases, and/or transmit to the concerned organizations.

5. Update Justification and/or Verification matrices, **and/or traceability matrices.**

D. Control the verification process. This activity includes the following tasks:

1. Coordinate the activities of the verification process with the person responsible for the activities of the system definition and integration processes (specifiers, architects, designers, developers, integrators) for the technical aspects related to the object submitted to verification (references, errors, defects, etc.).

2. Coordinate the activities of the verification process with the System Development Project Manager for the time and cost aspects of the execution of the verification actions or procedures (in particular for reviews, inspections, tests), for the acquisition of verification tools and enabling systems required for verification, for the availability of qualified personnel, for the procurement of resources, etc.

3. Coordinate the activities of the verification process with the Configuration Manager for the verification configurations, the anomaly or non-conformance sheets, the versions of references and of the objects to be verified.

4. Update the Verification Plan (verification strategy, verification actions, detailed calendars, etc.) as the project progresses.

5. Update the Verification procedures and all other items following anomalies found during the verifications execution.

5.2.1.4 ONTOLOGY ELEMENTS

Elements

The system verification process uses the main engineering meta-data indicated in Table 8, in alphabetical order.

Chapter 5.2.1. Verification of the system: Process approach

ELEMENT	DEFINITION AND ATTRIBUTES (examples)
Design property	A (quantitative) measure or an estimate of an architectural characteristic. It is associated with a system component, a physical interface, a physical architecture. It is a property obtained when defining the components of the system via the allocation of non-functional requirements, or obtained by means of techniques (estimation, analysis, study, calculation, simulation) with respect to a specific aspect, or obtained by definition in the case of an existing component.
	Identifier; Description; Comment
Engineering entity	Engineering meta-data: Stakeholder Requirement, System Requirement, Scenario (for logical architecture), Function, Flow, Physical architecture, Component, Physical interface, etc.
	Identifier; Description; Comment
Implemented component (also integrated system)	Component (System element) that has been implemented (manufactured, assembled); it is identified by a serial number. Examples: software application, mechanical / electrical / electronic hardware component, ... operator role, procedure, protocol, etc.
	Identifier; Description; Comment
System requirement	Statement that identifies an expected characteristic of a product, service, enterprise or process operational, functional, or constraint, which is unambiguous, testable or measurable, and necessary for product, service, enterprise or process acceptability. (Adapted from ISO/IEC 26702 = ISO/IEC 24748-4)
	Identifier; Description; Type; Comment
Verification action	A verification action describes what must be checked and how: the object submitted to verification, the expected result of the execution of this action, the method or technique to be applied, at which system decomposition level the action must be executed.
	Identifier; Description; Comment
Verification configuration	A verification configuration gathers physical elements (objects to be verified, verification tools or means) necessary for the execution of a Verification procedure.
	Identifier; Description; Comment
Verification procedure	A verification procedure groups a set of Verification actions executed together simultaneously or sequentially in a defined sequence within a Verification configuration.
	Identifier; Description; Duration; Time unit; Comment
Verification tool	Physical equipment and/or computer equipment used to perform one or more verification procedures (test bench, simulator, harness, cap, launcher, measurement device, enabling system, etc.).
	Identifier; Description; Comment

Table 8 - Main engineering meta-data related to verification

Chapter 5.2.1. Verification of the system: Process approach

Relationships

The main relationships between the engineering meta-data are presented in Figure 45.

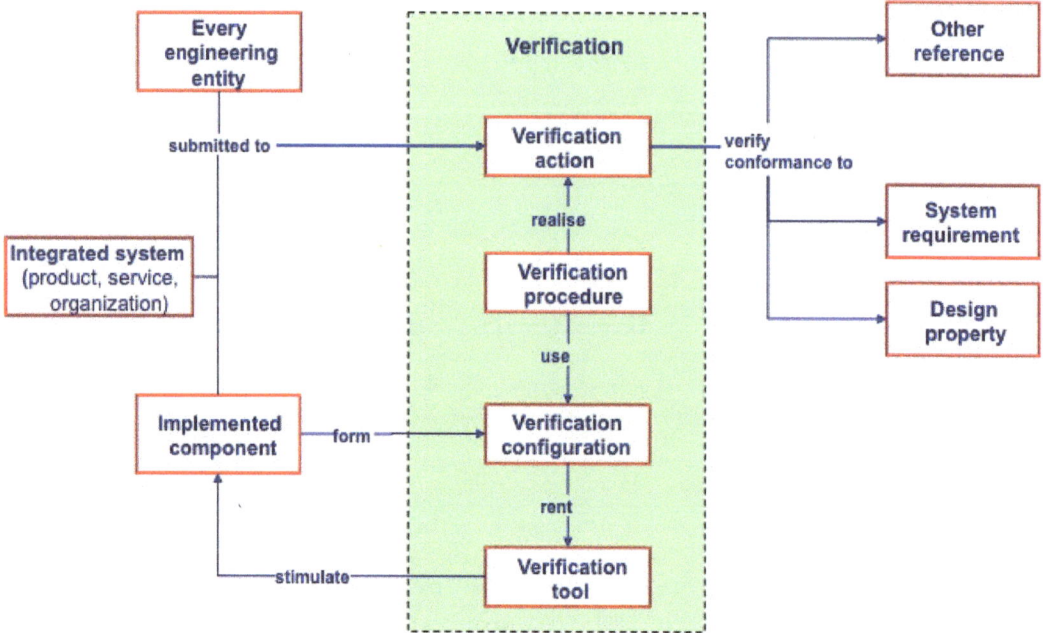

Figure 45 - Relationships between the engineering meta-data related to verification

Utilisation of ontology elements

The meta-data defined above in Table 8 and their relationships in Figure 45 represent the skeleton of the activities and tasks of the Verification Process. During the execution of tasks of the process the experts in systems engineering have to:

- Create as many as necessary instances of these meta-data: Verification action, Verification configuration, Verification procedure, Verification tool.
- Fill in attributes of these instances with values; the generic attributes represent characteristics of a meta-data; particular values for one instance particularise this instance among others.
- Instantiate the generic relationships between meta-data to establish traceability links between instances.

The main relationships to establish during the execution of this process are the following:

- a Verification action submits an Engineering entity
- an Engineering entity is submitted to a Verification action
- a Verification action submits an Implemented component / integrated system
- an Implemented component / integrated system is submitted to a Verification action
- a Verification action verifies the conformance to a System requirement, a Design property

- (the conformance to) a **System requirement**, a **Design property** is **verified by** a **Verification action**
- a **Verification action** is **realised by** a **Verification procedure**
- a **Verification procedure realises** a **Verification action**
- a **Verification procedure uses** a **Verification configuration**
- a **Verification configuration** is **used by** a **Verification procedure**
- a **Verification configuration rents** a **Verification tool**
- a **Verification tool** is **rent by** a **Verification configuration**
- a **Verification tool stimulates** an **Implemented component**
- an **Implemented component** / integrated system is **stimulated by** a **Verification tool**
- an **Implemented component** / integrated system **forms** a **Verification configuration**
- a **Verification configuration** is **formed by** an **Implemented component**

Implemented by software applications such as Data Bases or Repositories or Spread-sheet tools, these relationships enable the generation of Traceability Matrices; these latter are useful for various purposes such as verification, impact analysis (in case of evolutions), justification, technical risks management, and consistency checking.

5.2.1.5 ARTEFACTS - DOCUMENTATION

This process generates the following documents. Templates and Guidelines are provided in Annex.

- Verification Plan - see Annex chapter 8.2
- Verification Action Sheet - see Annex chapter 8.4
- Verification Procedure Sheet - see Annex chapter 8.5
- Verification Analysis Report
- Anomaly or Non-conformance Sheet - see Annex chapter 8.6

Chapter 5.2.2. Validation of the system: Process approach

5.2.2 Validation process

This section describes:

- ♦ The location of the process in the development cycle - section 5.2.2.1
- ♦ The purpose, inputs and outputs of the process - section 5.2.2.2
- ♦ The activities of the process - section 5.2.2.3
- ♦ The ontology elements used - section 5.2.2.4
- ♦ The main artefacts produced by the process - section 5.2.2.5

5.2.2.1 LOCATION OF THE VALIDATION PROCESS

Figure 46 shows the location of the validation process among the processes of definition, implementation and integration of the system. In the left part of the figure, the validation process covers definition elements, entities and artefacts. In the right part, it focuses on physical elements, entities and integration artefacts.

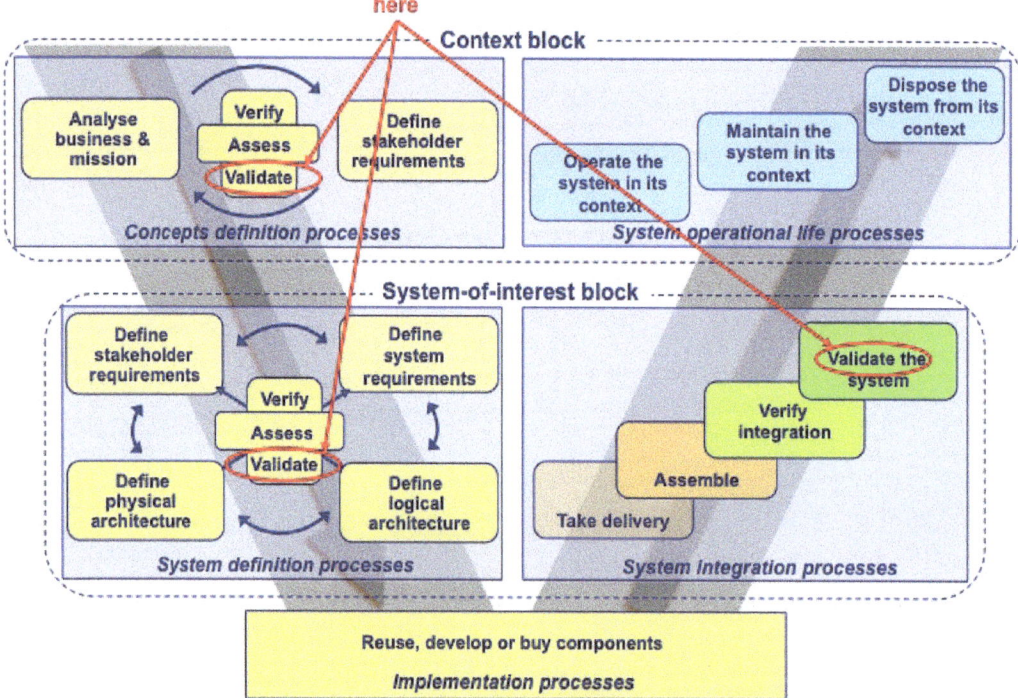

Figure 46 - Location of the validation process in the development

5.2.2.2 PURPOSE, INPUTS AND OUTPUTS OF THE PROCESS

Synthetic and generic expression

The purpose of the validation process is to determine that an implemented object* conforms with its requirements.

The aim of the validation is to gain confidence in the ability of the object to achieve its mission or function under specified conditions of use.

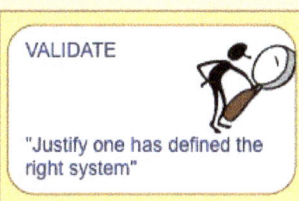

VALIDATE

"Justify one has defined the right system"

Discussion

* : By "implemented object", understand any engineering entity such as implemented component (product, service, organization), procedure, document, stakeholder requirement, system requirement, logical architecture, physical architecture, design properties, test procedure, etc.

The validation process is instantiated as many times as necessary during the development of the system and for each implemented object.

The validation process is used to determine that the transformation used to obtain each object has produced the expected results from provided inputs. Any validation is based on tangible evidence; that is to say on information whose veracity can be demonstrated, based on facts obtained using techniques, such as inspection, test, analysis, demonstration, simulation, etc. - see section 5.1.3.

To validate a system (product, service, organization) is to determine that it conforms with its System Requirements and possibly its Stakeholder Requirements, depending on the agreed contract. It is in this sense that the following descriptions are given, i.e., the validation process is instantiated for the definition and integration of the concerned system.

The validation process is therefore not limited to a stage or phase at the end of the development of the system. It must be used for all engineering data and for any intermediate results; so we will start at the beginning of the life cycle by validating operational concepts and scenarios, validating Stakeholder Requirements, ... well before thinking about validating the complete system.

The validation process applied to the fully integrated system is often referred to as Final Validation. In certain industrial areas, there is a factory validation (the reference used consists of the System Requirements), and on site or in operation validation, sometimes called operational validation or qualification (the reference used consists of Stakeholder Requirements).

Note: The standard ISO/IEC/IEEE 15288, System life cycle processes [ISO 15288], provides the following purpose of the system validation process; the expression given in this book is therefore broader.

The purpose of the Validation process is to provide objective evidence that the system, when in use, fulfils its business or mission objectives and stakeholder requirements, achieving its intended use in its intended operational environment.

The objective of validating a system or system element is to acquire confidence in its ability to achieve its intended mission, or use, under specific operational conditions. Validation is ratified by stakeholders. This process provides the necessary information so that identified anomalies can be resolved by the appropriate technical process where the anomaly was created.

Chapter 5.2.2. Validation of the system: Process approach

Inputs

The inputs of the process are:

- the object submitted to validation,
- the **definition references** of the object submitted to validation with respect to the specific characteristics of this object,
- the validation means and/or tools required to carry out the validation actions.

Notes:

To validate the system definition, each engineering data must be validated; for example: **Stakeholder Requirements**, **System Requirements**, logical architecture elements / entities (**Functions**, **Input / Output Flows**, etc.), physical architecture elements (**Component**, **Physical Interface**), **Design Properties**, and so on. The reader can do the exercise of finding the **definition references** of each of the engineering entities and deducing the specific characteristics to be validated. However, this information is indicated in the description of activities and in a section of each system definition process, in Volume 2, Systems Opportunities and Requirements [Faisandier2 2013] and in Volume 3, Systems Architecture and Design [Faisandier3 2013], of this series of books.

To validate the integration of the system, i.e. the implemented product, service or organization, the system **definition references** essentially consist of the System Requirements Document and the Stakeholder Requirements Document.

Outputs

The outputs of the process are:

- the validation plan that includes:
    - the validation strategy, or the reference to the system verification - validation strategy
    - the definition of the selected **Validation actions**
- the **Validation procedures**
- the **Validation configurations**
- the needs and/or the requirements for the **Validation tools**
- the results of the execution of the **Validation actions**
- the reports containing the analysis of the results of validations
- the anomaly and/or non-conformance sheets

Note: The overall result of the correct execution of the process is a validated system (product, service, organization); it is an observation (validated system) and not a transformation.

5.2.2.3 ACTIVITIES OF THE PROCESS

The activities of the validation process follow:

      A. Prepare the validation

      B. Perform the validation

      C. Analyse and record the results of the validation

      D. Control the validation process

Figure 47 presents the activities of the process, the exchanged engineering data and the main artefacts generated by the process.

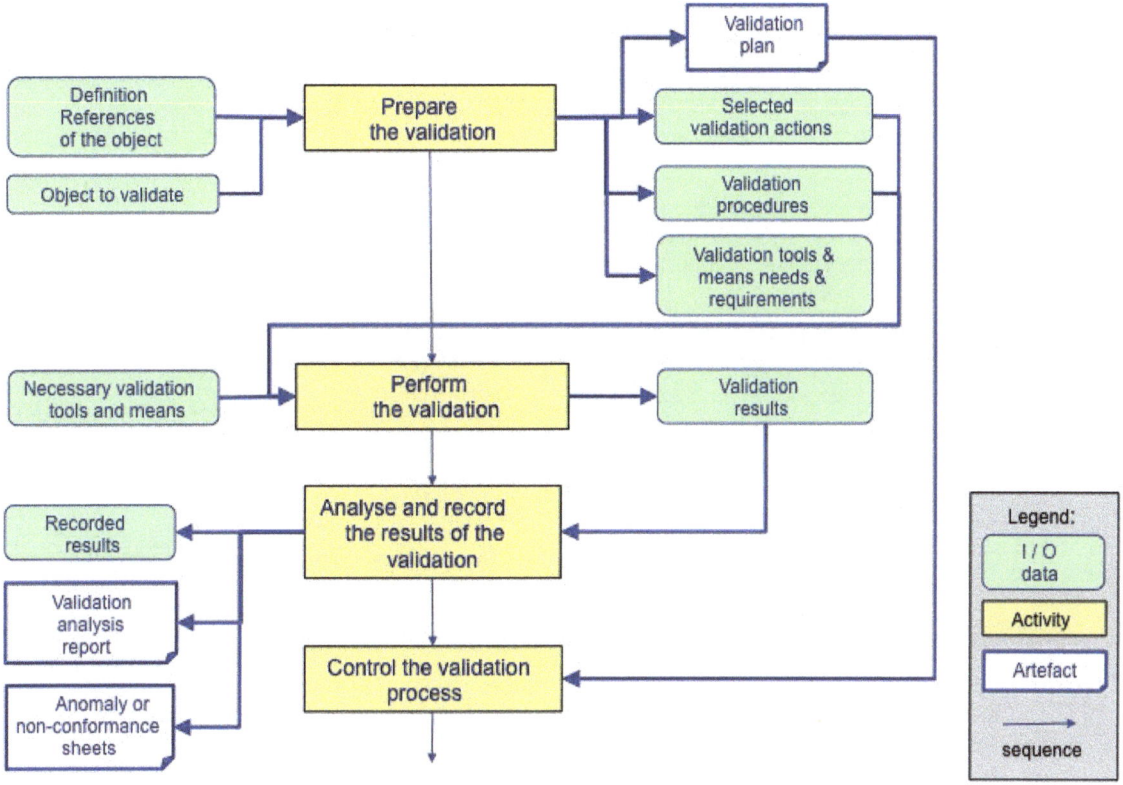

Figure 47 - Activities of the validation process

Chapter 5.2.2. Validation of the system: Process approach

Details of activities

A. **Prepare the validation.** This activity includes the following tasks:

1. Define the validation strategy. This task is composed of the following sub-tasks:

- Analyse the perimeter of the validation - see section 5.1.5. This analysis leads to the definition of a list of objects to be validated and of **Validation actions** to be exercised on these objects - see section 5.1.2. This analysis is normally carried out as soon as the concerned objects are defined. For example, as soon as a **Stakeholder Requirement** is written or expressed, define **Validation actions** to detect whether the requirement as written is *validable*. If this is not the case, just after the execution of the Validation action, it will be necessary to reformulate or re-examine the requirement, or to break it down into several requirements; it is the same with the **System Requirements**, the elements and characteristics of logical and physical architecture, and also with the **Design properties**. In this sense, the validation process exerts a significant feedback on engineering if it is executed early and iteratively.

- Define qualitative and quantitative objectives (coverage rates - see section 5.1.5) for each object to be validated. This allows providing criteria to stop the validations.

- Establish the list of limiting or imposed constraints - see section 5.1.5. These constraints impact the implementation of **Validation actions** and reduce the number of **Validation actions**; example of constraints: feasibility, cost, deadline, means, safety, regulations, accessibility and measurability.

- Select the validation techniques or methods relating to the objects to be validated and the **Validation actions** taking into account limiting or imposed constraints - see section 5.1.3 (for example: demonstration, inspection, tests, simulation, etc.).

- Select the **Validation tools** or means required to perform the **Validation actions** and the associated validation techniques or methods; they are physical means (test bench, mock-up, measurement devices, etc.), qualified personnel, consumables, infrastructure, etc.

- Define the scope of the validation by comparing validation objectives, limiting constraints and the means required for validations. This leads to the definition of a list of **selected Validation actions**, at what times it is appropriate to execute them, and by what means. Nevertheless, a risk analysis should be conducted to identify unacceptable or acceptable risks if a **Validation action** is removed from this list.

- Organize the **Validation actions** by dividing them into the various phases of development, the different levels of decomposition (system, subsystem, component), different validation techniques or methods, different **Validation tools** or means. Practically, traceability matrices (allocation matrices and justification matrices) can be used or consolidated.

- Optimize the **validation strategy** against the project deadlines and costs and the **Validation tools** or means needed, by analysing the verification strategy and the integration strategy to detect synergies. Delete redundant **Validation actions**; if possible, group **Verification actions** and **Validation actions**; postpone the execution of certain **Validation actions** from a system level to a higher or lower level to avoid developing expensive **Validation tools** or means. Promote **Validation actions** that can detect the most potential defects, and/or to gain the best confidence.

- Document the **Validation actions** in Validation Action Sheets - see section 5.2.2.5, and the validation strategy in the System Validation Plan - see section 5.2.2.5.

2. Develop the validation procedures. A **Validation procedure** generally groups several **Validation actions** on the same **Validation tool** or means. It refers to the **Validation actions** to be performed, describes the necessary **Validation configuration** (objects to be validated, initial conditions, tools, personnel skills), execution steps and stimuli, checkpoints and results to be recorded. Document in Validation Procedure Sheets - see section 5.2.2.5.

3. Identify the constraints induced by the validations on the system-of-interest and its Components. These constraints are revealed when the validation strategy is established; they derive from the study of the feasibility or the implementation of certain selected validation actions. These constraints can give rise to **System requirements**, architectural arrangements (**Component, Physical Interface**, volume, surface, location, etc.), design layouts, and **Stakeholder requirements**. They must be provided to the concerned processes in order to be incorporated in the corresponding references. These constraints concern, for example, performance or effectiveness characteristics to be validated, accessibility to certain components or interfaces, measures to be carried out, etc.

4. Define or specify the **Validation tools** or means necessary for the execution of the **Validation actions** and their deadlines for availability. These means can be tools, products, services and/or enabling systems. This task includes defining the technical requirements of these enabling elements, their interfaces with the system-of-interest, and the deadlines for delivery or availability.

5. Obtain the **Validation tools** or means. The means must be available at the appropriate time to be able to perform the **Validation actions** using the previously defined **Validation procedures**. The means can be obtained in different ways, by leasing, acquisition, development, reuse, subcontracting. One or more means of validation can be the object of a complete enabling system developed in the context of a project separate from that of the system-of-interest. Attention, the development of some means may be longer than that of the system-of-interest; it is a matter of coordinating developments over time.

B. Perform the validation. This activity consists of carrying out the basic operations described in the Validation Plan and in the **Validation actions** using the **Validation Procedures**, with the means and the **Validation tools** provided, under the prescribed conditions and in the forecast calendar. More precisely:

1. Gather the objects to be validated into subassemblies as provided for in the **Validation procedures**.

2. Construct the **Validation configurations** (objects to be validated, means and tools, qualified personnel, resources or consumables required) and set up the conditions stipulated in the **Validation procedures**.

3. Perform the **Validation procedures** as prescribed, following the step-by-step procedure.

4. Collect the results obtained through the execution of the **Validation procedures** and the intermediate measures required by the **Validation actions**. Note the gaps between the defined conditions and the actual conditions of execution.

C. Analyse and record the results of validation. This activity consists of providing diagnostics and recommendations. The tasks are the following:

1. Compare obtained results against the expected results as described in the **Validation actions**.

    - If there is no gap, the object conforms with the expected characteristic.
    - If there is a gap, decide on the acceptability of this gap. If the gap is accepted, the object conforms; justify the acceptance. If the gap is not accepted, the object does not conform with the expected characteristic.

2. Record the status of the comparison in the validation Analysis Reports or Sheets provided for this purpose: conform and rationale in case of gap, or non-conform. In the non-conform case, an Anomaly or Non-conformance Sheet is initialized.

3. Investigate the cause(s) of the gap in the case of non-conformance. The causes can be: an error in the validated object, an error in the **Validation action** or in its execution, an unrealistic

expected result (the references are involved), an error in the **Validation procedure**, in the **Validation configuration**, in the execution conditions, etc. If the validated object or its references are involved, any changes must be made by the concerned processes; under no circumstances by the validation process. All other cases are the responsibility of the validation process, which must modify the concerned elements, before being able to be used again in subsequent validation sessions.

4. Establish the Validations Analysis Reports and any Anomaly or Non-conformance Sheets; record the information in the project databases, and/or transmit to the concerned organizations.

5. Update Justification and/or Verification matrices, and/or traceability matrices.

D. Control the validation process. This activity includes following the tasks:

1. Coordinate the activities of the validation process with the person responsible for the activities of the system definition and integration processes (customer or users representatives, specifiers, architects, designers, developers, integrators) for the technical aspects related to the object submitted to validation (references, errors, defects, etc.). In particular, with regard to the acceptability of any gaps.

2. Coordinate the activities of the validation process with the System Development Project Manager for the time and cost aspects of the execution of the validation actions or procedures (in particular for reviews, inspections, tests), for the acquisition of validation tools and enabling systems required for validation, for the availability of qualified personnel, for the procurement of resources, etc.

3. Coordinate the activities of the validation process with the Configuration Manager for the validation configurations, the anomaly or non-conformance sheets, the versions of references and of the objects to be validated.

4. Update the Validation Plan (validation strategy, Validation actions, detailed calendars, etc.) as the project progresses.

5. Update the Validation procedures and all other items following anomalies found during the validations execution.

5.2.2.4 ONTOLOGY ELEMENTS

Elements

The system validation process uses the main engineering meta-data indicated in Table 9, in alphabetical order.

Chapter 5.2.2. Validation of the system: Process approach

ELEMENT	DEFINITION AND ATTRIBUTES (examples)
Engineering entity	Engineering meta-data: Stakeholder Requirement, System Requirement, Scenario (for logical architecture), Function, Flow, Physical architecture, Component, Physical interface, etc.
	Identifier; Description; Comment
Implemented component (also integrated system)	Component (System element) that has been implemented (manufactured, assembled); it is identified by a serial number. Examples: software application, mechanical / electrical / electronic hardware component, ... operator role, procedure, protocol, etc.
	Identifier; Description; Comment
Stakeholder requirement	Necessity or desire expected by an end user, formally drafted and expressed in terms of a client, of an end user; service, objective, capability expected from the future system by the end users. Equivalent to expectation; includes user requirements.
	Identifier; Description; Type; Comment
System requirement	Statement that identifies an expected characteristic of a product, service, enterprise or process operational, functional, or constraint, which is unambiguous, testable or measurable, and necessary for product, service, enterprise or process acceptability. (Adapted from ISO/IEC 26702 = ISO/IEC 24748-4)
	Identifier; Description; Type; Comment
Validation action	A validation action describes what must be checked and how: the object submitted to validation, the expected result of the execution of this action, the method or technique to be applied, at which system decomposition level the action must be executed.
	Identifier; Description; Comment
Validation configuration	A validation configuration gathers physical elements (objects to be validated, validation tools or means) necessary for the execution of a Validation procedure.
	Identifier; Description; Comment
Validation procedure	A validation procedure groups a set of Validation actions executed together simultaneously or sequentially in a defined sequence within a Validation configuration.
	Identifier; Description; Duration; Time unit; Comment
Validation tool	Physical equipment and/or computer equipment used to perform one or more validation procedures (test bench, simulator, harness, cap, launcher, measurement device, enabling system, etc.).
	Identifier; Description; Comment

Table 9 - Main engineering meta-data related to validation

Chapter 5.2.2. Validation of the system: Process approach

Relationships

The main relationships between the engineering meta-data are presented in Figure 48.

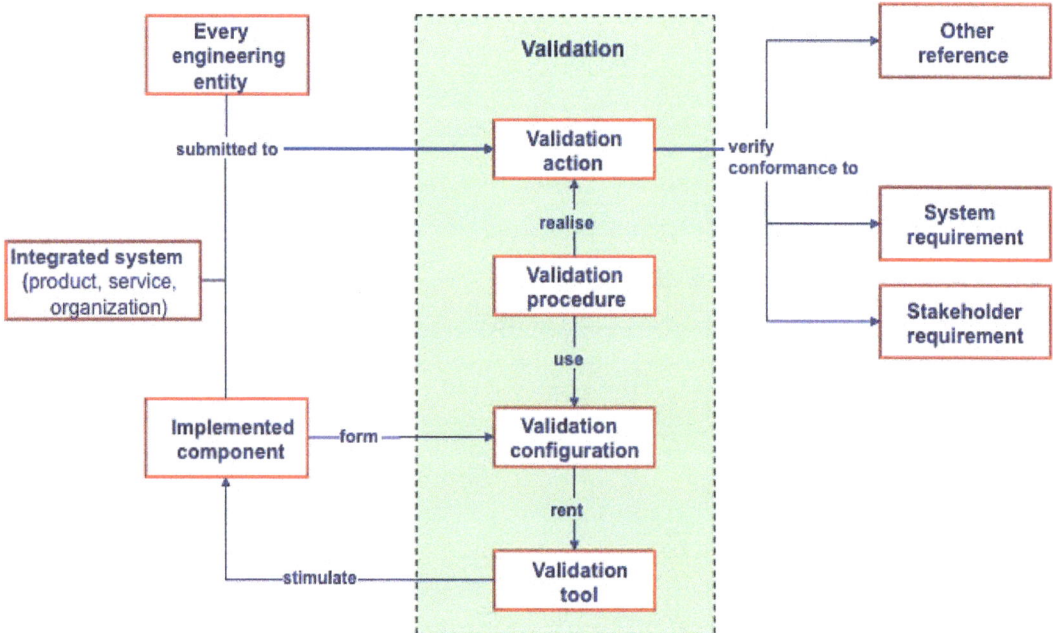

Figure 48 - Relationships between the engineering meta-data related to validation

Utilisation of ontology elements

The meta-data defined above in Table 9 and their relationships in Figure 48 represent the skeleton of the activities and tasks of the Validation Process. During the execution of tasks of the process the experts in systems engineering have to:

- Create as many as necessary instances of these meta-data: Validation action, Validation configuration, Validation procedure, Validation tool.

- Fill in attributes of these instances with values; the generic attributes represent characteristics of a meta-data; particular values for one instance particularise this instance among others.

- Instantiate the generic relationships between meta-data to establish traceability links between instances.

The main relationships to establish during the execution of this process are the following:

- a Validation action submits an Engineering entity
- an Engineering entity is submitted to a Validation action
- a Validation action submits an Implemented component / integrated system
- an Implemented component / integrated system is submitted to a Validation action
- a Validation action verifies the conformance to a System requirement, a Stakeholder requirement

148

- (the conformance to) a **System requirement**, a **Stakeholder requirement** is **verified by** a **Validation action**
- a **Validation action** is **realised by** a **Validation procedure**
- a **Validation procedure** **realises** a **Validation action**
- a **Validation procedure** **uses** a **Validation configuration**
- a **Validation configuration** is **used by** a **Validation procedure**
- a **Validation configuration** **rents** a **Validation tool**
- a **Validation tool** is **rent by** a **Validation configuration**
- a **Validation tool** **stimulates** an **Implemented component (or system)**
- an **Implemented component (or system)** is **stimulated by** a **Validation tool**
- an **Implemented component (or system)** **forms** a **Validation configuration**
- a **Validation configuration** is **formed by** an **Implemented component (or system)**

Implemented by software applications such as Data Bases or Repositories or Spread-sheet tools, these relationships enable the generation of Traceability Matrices; these latter are useful for various purposes such as validation, impact analysis (in case of evolutions), justification, technical risks management, and consistency checking.

5.2.2.5 ARTEFACTS - DOCUMENTATION

This process generates the following documents. Templates and Guidelines are provided in Annex.

- Validation Plan - see Annex chapter 8.2
- Justification Matrix - see Annex chapter 8.3
- Validation Action Sheet - see Annex chapter 8.4
- Validation Procedure Sheet - see Annex chapter 8.5
- Validation Analysis Report
- Anomaly or Non-conformance Sheet - see Annex chapter 8.6

5.3 Practice

This chapter provides explanations, practices and recommendations related to verification and validation:

- Traceability and justification matrix - section 5.3.1
- Inspection technique - section 5.3.2
- End-of-activity review technique - section 5.3.3
- Test technique - section 5.3.4
- Organization of verification and validation activities, responsibility - section 5.3.5
- Recommendations and FAQ - section 5.3.6

5.3.1 Traceability and justification matrix

Traceability

The set of relations established throughout the development between all the definition entities instances (essentially: stakeholder requirements, system requirements, functions, flows, components, physical interfaces), then with integration entities (implemented components, integration aggregates), forms what is called **traceability**. Traceability is not an engineering method, but it contributes greatly to the verification of the consistency and the exhaustiveness of the engineering, and also to the validation of the engineering.

During the engineering, the following matrices are progressively established:

- The allocation matrix, or correspondence table between stakeholder requirements and system requirements, and the inverse matrix
- The allocation matrix, or correspondence table between system requirements and functional architecture elements (functions, input-output flows, operational modes, transitions of mode, scenarios), and the inverse matrix
- The allocation matrix, or correspondence table between system requirements and physical architecture elements (components, physical interfaces, architectures), and the inverse matrix
- The allocation matrix of functions to components, and the inverse matrix
- The allocation matrix of input-output flows to physical interfaces, and the inverse matrix

These matrices, or tables, are part of the rationale for engineering, since they allow questions to be answered: What is the origin of such engineering data? Are there any orphan (forgotten), or redundant data? Is there consistency in the traceability paths that link the allocation matrices?

The answer to the other questions related to the justification of engineering (Why such an element of solution? and How is such an element obtained?) is provided by the choices and the pertinence of these choices through the system properties assessment process - see Chapter 4.

These matrices are particularly important for stakeholder requirements with respect to system requirements for final validation. If the translation and allocation of stakeholder requirements into system requirements (e.g. translation of an expected service into functional requirements, and the allocation of global objectives into effectiveness requirements) have been correctly established, the validation of the implementation of the system requirements in the final product should avoid defining and implementing many costly validation actions with regard to stakeholder requirements.

Justification Matrix

The implementation of the verification and validation processes is based on the development of the verification - validation strategy, see section 5.1.5. This strategy is represented concretely as a table, or a matrix, or a database (according to the number of headings and established relations). The Justification Matrix is also called the Verification Matrix. It describes how the implementation of each requirement and of each design property in the implemented system (product, service or organization) can be checked by means of verification-validation actions. These verification - validation actions can be theoretical (they concern the definition of the system - its engineering) or experimental (they concern the object realized).

The Justification Matrix is a tool to help validate compliance with requirements and design properties by the implemented system (product, service or organization). For each requirement and for each property to be verified and/or validated; it contains the following headings:

- The verification or validation action to be carried out in order to check compliance or non-compliance with the requirement or property, see section 5.1.2
- The object submitted to this verification or validation action
- The verification - validation techniques or methods to be used, see section 5.1.3
- The rationale for non-coverage (reason for the removal of the verification), if it is the case
- The phase of the life cycle during which the verification - validation action is performed; see Note below
- The level of decomposition of the system to which the verification - validation action is performed; see Note below
- The reference of the verification - validation Action Sheet
- The reference of the verification - validation configuration
- The progress of the verification - validation action (defined, in progress, performed)
- The verification - validation result synthesis (accepted, not accepted)
- The reference of the verification - validation Analysis Report

Note: From an optimization perspective (for example to reduce costs, or deadlines, or the use of means or tools), it is quite possible to:

- Postpone the execution of certain verification - validation actions upstream or downstream of the normally scheduled phase.
- Move the execution of some verification - validation actions to higher or lower level system blocks than the current system block. To do this, insert the requirements / properties of the related system blocks into the Justification Matrix and treat them as requirements / properties of the current system block.

The establishment of the Justification Matrix is an iterative work, which begins as soon as the system requirements are written; it continues throughout the engineering and integration work, and even beyond if compliance with stakeholder requirements or system requirements cannot be validated before the system is transferred into operation. The headings are filled in through the execution of the verification and validation processes activities.

To establish the Justification Matrix, it is recommended to start with the functional requirements, then the functional and physical interface requirements, then the requirements for operational modes, scenarios, environment, maintenance, storage, constraints; continue with effectiveness and justification requirements; finally finish with the design properties.

Figure 49 illustrates the principle of establishing and filling in the Justification Matrix. Annex chapter 8.3 provides implementation details, and an example.

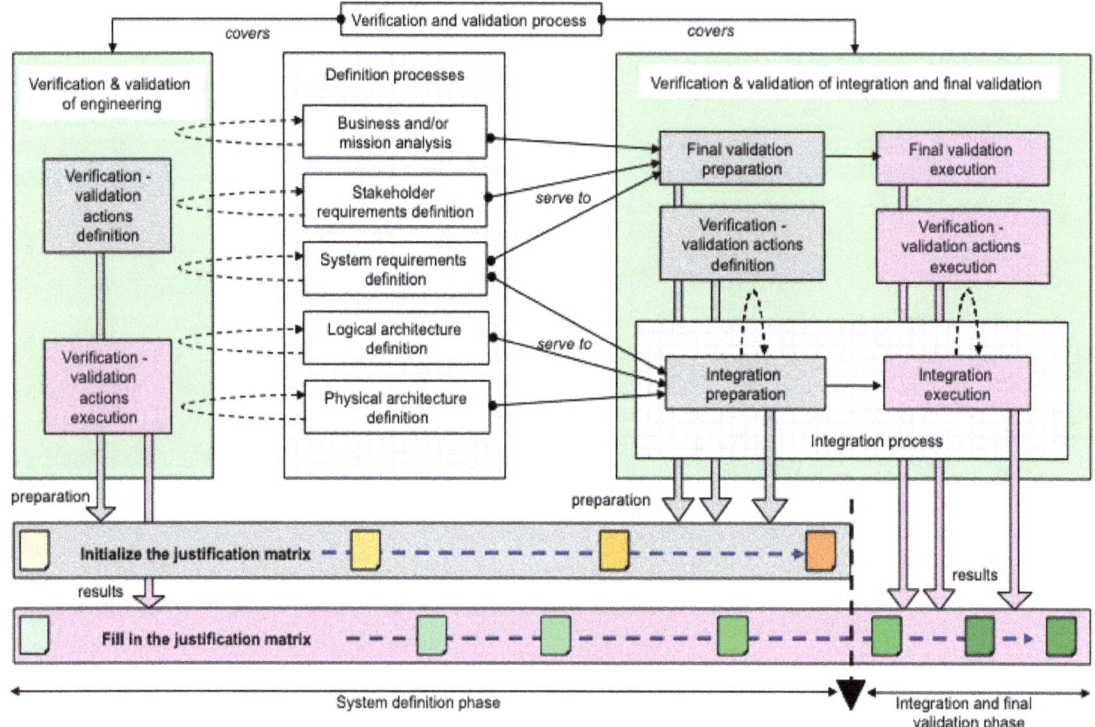

Figure 49 - Principle for establishing the Justification Matrix

Examples of verification - validation actions

These examples come from life-cycle processes as described in Volume 2, Systems Opportunities and Requirements [Faisandier2 2013], and Volume 3, Systems Architecture and Design [Faisandier3 2013], of this series of books. In the following lists, the distinction between verification action and validation action is not systematically made; the two types of action may reside in the same expression. This is an excellent exercise for the reader to determine in the following expressions, what relates to verification and/or validation.

Verification - validation actions related to the business or mission analysis process

The main elements to be verified and/or validated are the following:

- The identified issue, problem or opportunity is correctly expressed and understood by concerned stakeholders, and corresponds to a contextual situation.
- The main stakeholders have been identified and interviewed.

- Several potential solutions were examined, modelled, analysed, evaluated for feasibility and added value; they have been reviewed by stakeholders and concerned experts.
- The purpose, mission and objectives of a potential system have been formulated and are relevant to the contextual situation.
- Situations in the life of the potential system, as well as operational modes (states) have been identified and defined.
- Context relationships diagrams, which include the system and the physical elements of the context, have been developed and Services were deduced from these relationships.
- Context relationships diagrams, which include the system and stakeholders, have been developed and Constraints have been deduced from these relationships.
- Operational scenarios and/or use cases have been modelled and validated by experts and the stakeholders.
- A functional architecture of the context was modelled, and the functional interfaces were identified.
- A physical architecture of the context was modelled, and the physical interfaces were identified.
- Transitions between operational modes have been identified and modelled.
- Candidate solutions concepts (operational concepts and/or technological concepts) that support the system, in particular operational scenarios, and corresponding market strategies and/or business models have been devised, examined and evaluated.
- A trade-off study using assessment criteria was carried out to select the best solution concept, including the associated market strategy and business model.
- The selected solution concept, the market strategy and the associated business model have been reviewed and validated by the stakeholders.
- Feedback to and from other processes has been taken into account.

Verification - validation actions related to the stakeholder needs and requirements definition process

- Each characteristic of each stakeholder requirement, or set of stakeholder requirements, must be verified. In particular:
  - **Maturity**: Is the expression of stakeholder requirements close to the perceived needs or expectations of the stakeholders?
  - **Completeness**: Have all stakeholders been identified and interviewed? Are all stakeholder requirements expressed and written?
  - **Accuracy**: Do stakeholders recognize an exact expression of their expectations?
  - **Feasibility**: Are operational concepts identified to assess their feasibility and their ability to solve the submitted problem, issue or opportunity?
  - **Consistency**: Are there conflicts between the different stakeholder requirements expressed?

- The value of content, relevance, and meaning of stakeholder requirements must be validated. In particular:
  - **Understanding**: Validate understanding of stakeholder requirements with corresponding owners; this includes confirmation that the stakeholder requirements are understandable to the initiators and that resolution of conflict in the requirements has not corrupted or compromised the intentions of the stakeholders.
  - **Relevance**: Does the expression of the requirements make it possible to define the relevance of a solution to the submitted problem, issue or opportunity?
  - **Rationale**: Why do these needs, expectations and stakeholder requirements exist? What is the risk or cause that could cause the needs, expectations and stakeholder requirements to disappear?

Verification - validation actions related to the system requirements definition process

- The following characteristics of each requirement must be verified:
  - **Necessary**: The requirement defines an essential capacity, characteristic, or constraint. If it is removed, the lack created cannot be met by other capabilities or characteristics or constraints of the system (product, service or organization).
  - **Independence of the implementation**: The requirement, while responding to what is necessary and sufficient for the system, avoids putting unnecessary constraints on the architecture or the design. The goal is to be independent of implementation: the requirement expresses what is required, not how the requirement should be satisfied.
  - **Unambiguous**: The requirement is expressed in such a way that it can be interpreted in one way. The requirement is stated simply and is easy to understand.
  - **Consistency**: The requirement is free from conflict with other requirements.
  - **Complete**: The stated requirement does not need any additional explanation, and it adequately describes the capacity and characteristics to meet stakeholder requirements. This is facilitated if the system requirement is quantified.
  - **Singular**: The expression of the requirement has only one requirement without the use of conjunctions.
  - **Feasible**: The requirement is technically feasible and compatible with the constraints of the system (e.g. cost, schedule, technical, legal, regulatory constraints).
  - **Traceable**: The requirement is traceable upstream with specific documented stakeholder requirements, or other sources (e.g. market studies or concepts studies). The requirement is also traceable downstream with architectural or design properties, and/or with specific requirements applicable to lower level components. This means that all relationships with the requirement are identified in such a way that the requirement is traced with its source, architecture and design, and implementation.
  - **Verifiable**: There is a complete and economically appropriate method or technique for demonstrating compliance with the requirement in the product, service or organization. Verifiability is facilitated when the requirement is quantified.
- The following characteristics of each set of requirements must be verified:
  - **Complete**: The set of system requirements does not need any additional explanation, because it contains everything that is relevant to the definition of the system. In addition, the set contains no TBD (to be defined), TBC (to be confirmed), or TBR (to be resolved) clause. The resolution of designations TBX can be iterative and temporal (an acceptable deadline is fixed for TBX clauses, determined by risks and dependencies).

- **Consistency**: The set of system requirements does not contain conflicting individual requirements. The requirements are not duplicated. The same term is used for the same element in all requirements (no synonym).
- **Procurable**: The complete set of requirements can be satisfied by a solution that is commercially available or achievable under life cycle constraints (e.g., cost, schedule, technical, legal, regulatory constraints).
- **Delimited**: The set of system requirements confines the solution envisaged in the identified field, without going beyond what is necessary to satisfy stakeholder requirements.

Verification - validation actions related to the logical architecture definition process

The main elements to be verified / validated when defining the logical architecture are as follows:
- Each functional and interface requirement corresponds to one or more functions.
- The outputs of the functions correspond to the inputs provided.
- Each function produces at least one output.
- Functions are triggered by control flows if necessary.
- The functions are sequenced in the correct order and synchronized with one another.
- The duration of the execution of the functions is in the interval of the values indicated in the effectiveness or performance requirements.
- Every operational scenario is taken into account.
- Simulation of logical architecture models (or scenarios of functions and of operational modes) is complete in all possible cases, and shows that the consumption of the input flows and the production of output flows are correctly dimensioned (when simulation of models is possible).

Verification - validation actions related to the physical architecture definition process

The main elements to be verified / validated when defining the physical architecture are as follows:
- Each component of the system performs one or more functions of the logical architecture.
- Each function has been allocated to a component of the system.
- Each input-output flow is transported through a physical interface.
- The components of the context of the system-of-interest are connected to the components of this system-of-interest with physical interfaces (Links / Connectors).
- The logical architecture is correctly projected onto the physical architecture, and the allocated logical architecture correctly reflects this projection.
- The physical architecture is implementable by mature or mastered industrial technologies.
- The physical architecture is equipped with architectural characteristics and corresponding design properties.

Verification - validation actions related to the system properties assessment process

The main elements of this process to be verified and validated are described in section 4.2.5.

5.3.2 Inspection technique

Definition

Inspection is a verification technique planned within a project's activities (normally as soon as the submitted object exists), which consists of searching for defects in the system definition references (Stakeholder Requirements Document, System Requirements Document, System Architecture Description, Design Description, Manufacturing File) and in the implemented system (product, service, organization).

The inspection is carried out by subject matter experts in the relevant technical domains.

Purpose and objectives

The **purpose** of inspection is to:

- identify defects in the object submitted to inspection,
- check that the object possesses the expected characteristics,
- check that the professional rules have been applied.

The **objectives** of inspection are to:

- detect as many defects as possible in the object submitted to inspection, to eliminate them as soon as possible by corrective action, and before any use, and preferably before implementation,
- detect configurations of the object that may cause difficulties during testing or subsequent use of the object,
- identify the difficulties in applying the concerned professional rules,
- define and implement preventive actions for the necessary training and motivation to avoid systematically introducing defects,
- improve the maintainability of the system via the inspected object.

Inspections cover all development stages; they make it possible to detect and thus eliminate a large number of defects generated almost at their source.

Actors and execution

The actors of an inspection are:

- developers or authors of the object,
- inspectors or readers.

An inspection is executed in four steps:

- The developers or authors provide the object to be inspected, the verification references, and what they expect from the inspection.
- The inspector or reader reports subject matter and form errors (inconsistency, duplication, redundancy, lack), and asks questions of understanding, structuring, business rules usage, and solution selection. He or she makes positive remarks and suggestions; his or her attitude is critical and constructive.
- The developers or authors record the reported defects and answer the questions, which leads them in general to modify and correct the implemented object.
- The developers or authors inform the inspector of the follow-up given to his or her remarks and justify the discrepancies in writing.

Figure 50 summarizes the execution of the inspection.

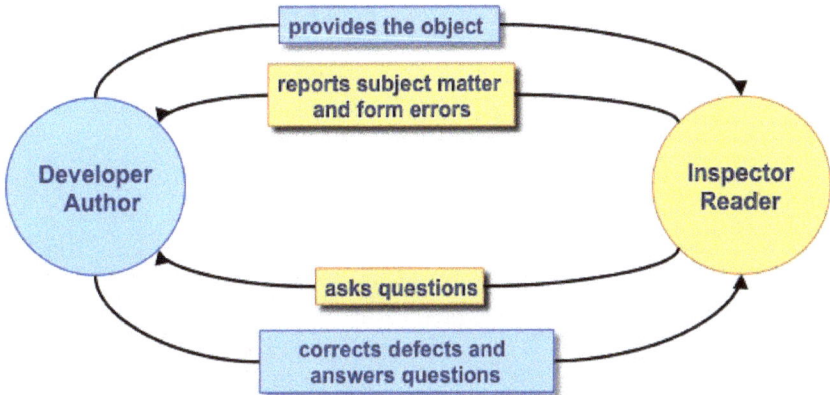

Figure 50 - Execution of the inspection

Effectiveness of the inspection

There are several types of inspections:

- Simple individual inspection with a developer and an inspector.
- Peer review by reciprocal exchanges of inspected objects between two developers of the same team or profession.
- Group inspection used on certain critical parts to reinforce the developer in the technical solutions adopted, or to solve technical difficulties encountered by the developer.

In order to carry out an effective inspection, the following points are decisive:

- Define precisely the input elements of the inspection: the verification references to be used, the Verification actions to be executed, and the input data.
- Use qualified inspectors in the relevant technology or profession, with a critical analytical capacity, available (planned activity) and motivated (personal involvement).
- Make checklists specific to the activity and the object inspected; these lists contain typical defects and are gradually enriched; they are a history of common mistakes.
- Draw up remarks and questions in the form of sheets, in order to keep track and facilitate their management; the written form obliges clear formulation and avoids inefficient discussions. See Inspection Sheet template in Annex chapter 8.7.
- The inspector must be informed as to whether or not his remarks are taken into account; obtaining relevant remarks is even better than the inspector's concern. If the developer does not take account of a remark made by the inspector, he or she must justify it.
- Inspection is outside the hierarchy; the only role of the hierarchical manager is to rule on the justifications for not taking into account the inspector's remarks.

5.3.3 End-of-activity review technique

Definition

The **end-of-activity review** is a validation technique, planned as part of the progress of a project, which consists of collecting subject matter expert opinions on the results of one or more activities and formulating recommendations.

The end-of-activity review is generally carried out during a milestone separating two phases of the development cycle. Its purpose is to decide on the results of the activities carried out during the phase that is ending, to close this phase and to authorize the start of activities in the next phase.

Purpose and objectives

The **purpose** of the end-of-activity review is to:

- decide on the closure of the corrective or preventive actions decided during the previous review,
- validate the technical elements developed during the phase which ends, in relation to the entry references,
- decide on the conformance and consistency of these technical elements with each other, in relation to inputs, and in relation to the standards, methods and tools used,
- check the availability and appropriate level of quality of the methods, tools, resources, timetables required for the activities of the next phase.

The **objectives** of the end-of-activity review are to:

- serve those responsible for the project to control and verify the consistency of the work entrusted to them,
- identify the current difficulties, the potential difficulties to come, and suggest corrections or improvements,
- determine the hard spots and estimate the workload to provide,
- give decision-makers arguments so that they can determine, at the appropriate time, their decision as to the continuation of the work.

Actors and execution

The actors of the end-of-activity review are:

- the **project team** (supplier, the people carrying out the concerned activity, who produce elements): it provides the elements subject to review, answers questions from the review group, applies the corrective actions or improvements decided by the steering committee,
- the **review group** made up of representatives of the client (acquirer) and experts from outside the project: it performs inspections and reviews of submitted elements, prepares problem sheets for the project team and prepares recommendations for the steering committee,
- the **steering committee** is made up of project managers on the client side (acquirer) and on the supplier side: it analyses the recommendations issued by the review group and decides on corrective actions or corresponding improvements according to the project parameters (risks, costs and deadlines impacted by the recommendations); the corrective actions are intended for the project team for application.

The end-of-activity review runs as follows:

1. The project team plans the scheduled review, producing the objects and/or documents submitted for review. The project manager looks for the chairman of the review group with the help of the management of the business unit. They jointly prepare the organizational note for the review, specifying the objectives of the review, members of the review group, meetings and their purpose, timing, tasks sharing, formats of problem sheets, recommendations and all necessary practical means.
2. Option: During a presentation meeting, the project team explains the purpose of the review, indicates the points on which it wishes opinions, its vision of the system, and potential risks.
3. The project team provides the review group with the elements submitted to review. The chairman of the review group states and explains the rules for the conduct of the review, and the deadlines.
4. The chairman of the review group distributes the elements and documents provided between the experts / members of the review group, indicates the instructions and specifies the schedule for any internal meetings between the members of the review group.
5. Each member of the review group critically examines the elements that are assigned to him / her, fill out Problem Sheets - see template in Annex chapter 8.8.
6. The secretary of the review centralises the Problem Sheets, classifies them, identifies them and sends them to the project manager.

Chapter 5.3. Verification and validation of the system: Practice

7. The project manager distributes the Problem Sheets among the project team members; the answers are prepared, and the Problem Sheets are completed. The Problem Sheets are sent to the secretary of the review who distributes them to the corresponding issuers.
8. The members of the review group read the answers and rule on: if agreed, the provision of the Problem Sheet becomes applicable and will be part of the recommendations or actions; otherwise the Problem Sheet is dealt with at a consultation meeting.
9. A consultation meeting is organized between the concerned members of the project team and the review group to discuss and clarify questions and answers.
10. The review group classifies the Problem Sheets against the types of provisions made and then draws up a synthesis report of the review giving its general and detailed recommendations.
11. The members of the steering committee study the synthesis report, transform or not, the recommendations and the provisions of the Problem Sheets into actions applicable to the project.
12. The steering committee provides to the project team the actions for implementation.

Figure 51 summarizes the execution of the end-of-activity review.

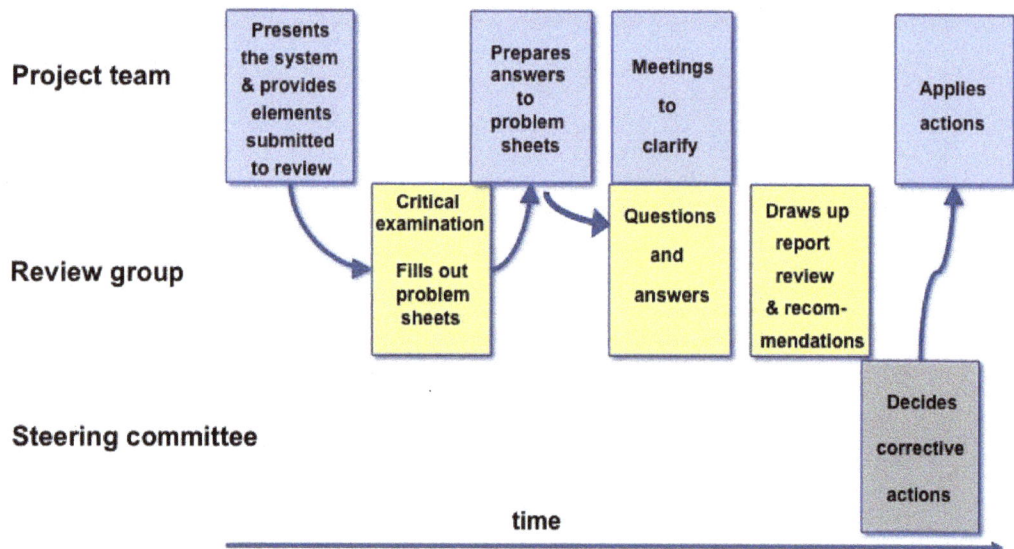

Figure 51 - Execution of the end-of-activity review

Effectiveness of the end-of-activity review

Within the framework of the projects, there are several types of reviews:

- **Joint reviews** that bring together the client and supplier. They relate to the system as a whole and authorize the crossing of the milestones defined at the beginning of the project, generally contractually.
- **Project reviews** to validate the results of activities carried out during a development phase and to authorize the start-up of activities in the next development phase.

To carry out an effective end-of-activity review, the following points are decisive:

- The end-of-activity review is a **validation** action, based on previous verifications. These verifications must be carried out preliminary to the validations subject of the review. In order to make the review more effective, it is therefore important to have carried out these verifications beforehand and to provide the members of the review with the findings and justifications.
- Identify as soon as possible the composition of the review group:
    - The chairman of the review group writes the organizational note of the review (objectives, assignment of tasks, schedule, list of submitted elements, documents) with the help of the project team manager, defines the agenda of meetings (difficulties, open issues, etc.). He / she must be trained with group leading techniques, have a review experience and must not be directly involved in the project in order to keep his / her independence of opinion.
    - Members of the review group are subject matter experts external to the project but of equivalent competence to the members of the project team; they treat each of their parts, they have a planned availability, they make positive remarks and suggestions; their attitude is to help the project team.
    - The secretary of the review is in charge of the practical organization of the review: calling to meetings, collecting of necessary documents, drafting and distribution of reports. Having documents well in advance allows review group members to prepare their job and thus be more effective.
- It is recommended to present the problems in the form of written sheets in order to keep track and to facilitate their management; the written form obliges clear formulation and avoids inefficient discussions.
- The project team classifies the Problem Sheets written by the review group into three categories - see Problem Sheet in Annex chapter 8.8:
    - sheets relating to form issues (avoid over-emphasis on this type of comment unless clarifying and avoiding confusion or misinterpretation),
    - sheets on issues and remarks resolved by a written answer from the project team,
    - sheets that require discussion between project team members and review group members (sometimes these are complex issues requiring data collection or preliminary analysis).
- Classify recommendations and actions:
    - Recommendations or minor actions of common sense and requiring no special workload. Once the decision is made, they do not require any specific follow-up.
    - Major recommendations or actions requiring special workload. Once the decision is made, they should be followed up at the project level.
    - Blocking recommendations or actions that condition the continuation of activities (without, however, requiring a further review after completion).
- Give final status to the end-of-activity review. This status can be:
    - positive, allowing for further action,
    - negative, requiring the resumption of a certain number of tasks and of supplementary reviews,
    - postponed, if the presented elements and/or evidences do not allow a ruling.
- Apply the decisions of the review. Once the recommendations have been accepted and decisions have been approved, their application by the project team must be checked.

Chapter 5.3. Verification and validation of the system: Practice

5.3.4 Test technique

5.3.4.1 NOTION OF TEST, PURPOSE, TYPOLOGY OF TESTS

Definitions

The **test** is a dynamic verification or validation technique that focuses on the working and behaviour of an **entity** over time.

The concerned entity may be a system, a component, a set of components, an integration aggregate, an object, a module, a function, an operation, a simulator, a process, etc. This entity performs internal functions (or actions) that transform input flows into output flows, possibly using its own resources.

The **working** of an entity is a sequence of internal functions that are triggered by external events (e.g. start and/or stop stimuli) under certain conditions or parameters (limits, dimensions, constraints, type, ranges, colour, etc.). Parameters determine input classes (for example, ranges of a voltmeter). Inputs, trigger events and constraints or parameters (input / output flows calibration) that constitute a set called **performing conditions** - see Figure 52.

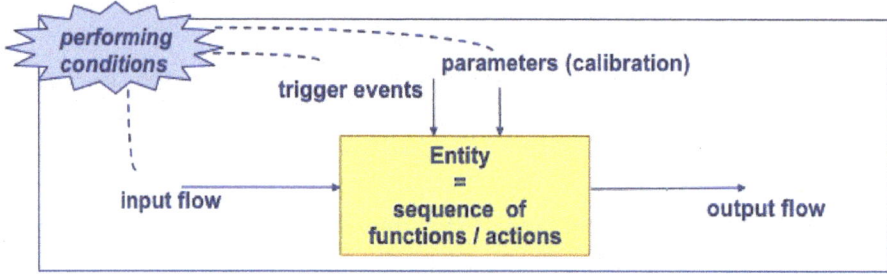

Figure 52 - Working of an entity

The **behaviour** of an entity is a chain of successive states caused by trigger events (external stimuli); the initial state of the entity (characterized by the initial performing conditions) and the final state (characterized by the results or output flows) are **observable** by the user of the entity. During a change of state and following internal functions or actions, the entity may pass through intermediate states (internal to the entity) observable or not observable by the user of the entity, see Figure 53.

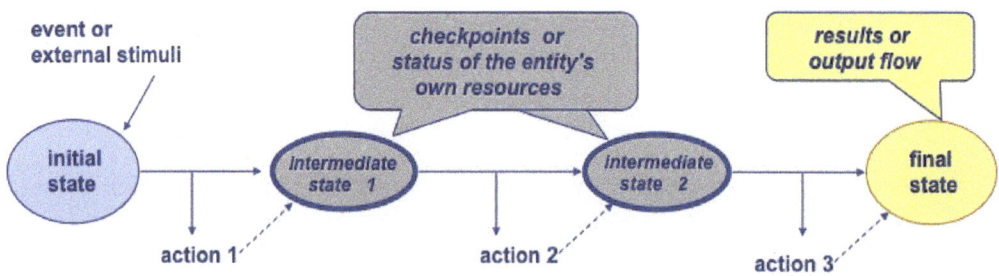

Figure 53 - Behaviour of an entity

Chapter 5.3. Verification and validation of the system: Practice

An entity is said **testable** if it contains items observable from outside. Observable items are either planned during the architecture and/or design (internal variables, counters, sensors, indicators, etc.) or added for testing purposes (instrumentation and check points).

A **test** is therefore an execution of the entity on a unique subset of the **performing conditions**, which makes the entity move from an **initial state** to a **final state** with possible intermediate state changes (internal), see Figure 54.

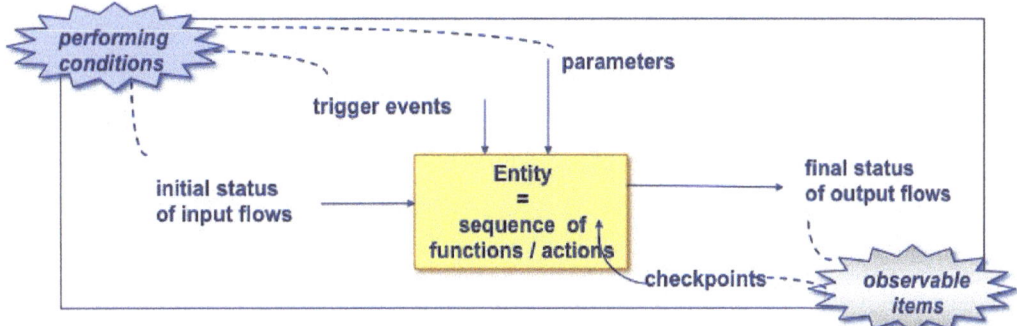

Figure 54 - Necessary elements for a generic test

When only the final state is observable, it is a **functional test** or a **black box** type test. Functional tests are only concerned with the result produced by the entity (output flow), and not how it produced it.

When the intermediate states are observable, it is a **structural test**, or a **glass box** type test or **grey box** type test. A structural test (strict sense) is concerned only with one particular intermediate state. A structural test uses the entity's own resources or check points.

A **test data set** is a **unique** combination of performing conditions, valued, characterized or quantified.

Purpose and objectives

The **purpose** of tests is to:

- show that the studied entity correctly performs the foreseen functions and effectiveness in the environment and the context which constrains it,
- establish conformance with regards to a reference, at least partially,
- investigate potential defects,
- establish a certain level of robustness.

The first two points address **conformance** or validation by demonstrating, with tests, that the entity satisfies its reference(s). The following two points address **defects searching** or verification by revealing, with tests, architectural, design or implementation defects.

The **objectives** of tests are to:

- verify the accuracy or not of certain assumptions made in architecture and design,
- better understand the behaviour of the concerned entity,

Chapter 5.3. Verification and validation of the system: Practice

- give a certain level of confidence in the working and behaviour of an entity in contexts close to the conditions of operational use,
- anticipate the entity's potential maintenance workload by using statistical data on defects searching efforts and their elimination.

Comprehension exercise

Requirements:

1. A vending machine distributes products at 30c and 40c.
2. The vending machine accepts 10c and 20c coins.
3. The vending machine gives the change.

Question:

What generic tests have to be defined to validate the implementation of requirements?

Method:

Identify actions or functions, input flows, output flows, triggers, parameters, own resources. Deduce the possible tests, the performing conditions and the expected output.

Solution:

- *Input flow*: coins
- *Output flow*: product and coins
- *Actions / functions*: accept (coins) - distribute (products) - give back (change)
- *Parameters*: currency
- *Triggers*: no specified stimuli, but it is possible to refer to existing vending machine: the introduction of coins or the choice of the product by the user.
- *Resources and own mechanisms*:
  - power supply (stand-alone or external network), products,
  - distribution mechanism, decision mechanism, products storage mechanism, coins storage mechanism (for entering and giving change).
- *Generic tests*:
  - With functional tests (black box type), we can validate only the distribution of products and the giving of change,
  - The requirements do not provide information on the processing of information. It is therefore necessary to consult the definition document to find the structural tests to be carried out. These structural tests can involve two actions: the choice of the product and the display intended for the user.
- *Performing conditions*: present coins (value greater than or equal to that of the product), ad hoc currency, current and correct choice.
- *Expected output*: product and currency.

Note: The detailed definition of the tests requires the use of test cases and test data sets development methods. See section further below.

Typology of tests

The tests can be classified into five categories: functional tests, glass box structural tests, grey box structural tests, environmental tests and effectiveness tests.

The **functional tests** are black box type tests; they concern the function(s) of the entity, that is to say the execution of the actions and transformations of the input flows into output flows. These tests focus only on the provided service, the outcomes (output flows), and not on the components of the entity to provide this service. This category includes:

- **Conformance tests**, which verify all expected actions specified or described in the definition references. These test cases are determined by selecting the combinations of the performing conditions which make it possible to reach the expected final state; these conditions must be described in the references: expected stimuli / triggers, limits, constraints and input flows.

- **Defects searching tests**, which are concerned with all other performing conditions (conditions not defined in the definition reference(s)).

Glass box type structural tests are concerned with the structure of the entity about undefined (unforeseen) behaviours in the requirements. These tests verify that the arrangements made in the architecture, design and/or implementation of the entity are correctly exercised and do not give rise to any malfunction. These are **defects searching tests** to reveal inaccessible parts (see Note 1) during nominal behaviour; they are interested, for example, in the transition between two successive intermediate states.

Structural tests are not conformance tests, because the provisions or arrangements checked are not expressed in the definition references (requirements, architecture, design). The developer has introduced, for prevention or precaution, or by business expertise, specific treatments, arrangements or provisions with regard to potential malfunctions (for example: bumpers, safety devices, defence type treatment, overflow, division by zero, etc.). This is the case, for example, for dependable systems or systems that have a high level of safety. These tests are therefore indispensable, because they explore how the concerned entity behaves, acts or reacts beyond the limits or thresholds. They typically require local insertion instrumentation.

Grey box type structural tests are concerned with the structure of the entity about foreseen non-nominal behaviours in the architecture or design (but not necessarily defined in the definition references), and the status of the entity's own resources. These tests verify that the arrangements made in the architecture, design and/or implementation of the entity are properly exercised. This category includes:

- **Conformance tests** that supplement the functional tests for cases that are difficult to access (see Note 1) and the cases of errors described in the definition references.

- **Defects searching tests** to reveal omissions or parts that are never accessible or not realized, not described in the definition references (see Note 2); they are interested in the successive intermediate states **and** the final state of the entity.

These tests involve particular cases, such as error cases, resources sharing, synchronization task or mechanism, recovery after failure, etc. They perform all the transitions between the successive intermediate states until the final state. They use specific instrumentation, and/or checkpoints, and/or entities own resources.

Note 1: Inaccessible parts come from functionalities existing in the entity, but intentionally unused. This is the case of software dead code introduced voluntarily (case of re-use) or unintentionally (following the omission of a decision, or of connection conditions, etc.). Dead codes must be identified, deleted, or blocked by a plug. The same thing happens in unused hardware devices present in reused equipment. In the case of dependable systems, the so-called non-accessible parts must be systematically searched for, and then treated by deletion, or by a stopper to prevent the solicitation.

Note 2: Unrealized parts are those that were forgotten during the implementation.

Environmental tests refer to the environmental conditions to which the entity is submitted. These tests often repeat tests of the previous categories, but focus on the different environmental or ambient conditions. We distinguish the following environmental or ambient conditions:

- mechanical: sinusoidal vibrations, random vibrations, acoustic vibrations, shocks, constant accelerations, blast, static deformations, platform movements
- climate: temperature, fire, humidity, solar flow, thermal shock, marine atmosphere, condensation, freezing and thawing, sea bundles, rain, snow, immersion, hail, ice, pressure, wind, mould, sand, dust
- electromagnetic: interference, electric fields, static electricity, lightning
- various: biological / bacteriological, corrosion by fluid and gas, chemical, nuclear radiation, radiological

Effectiveness tests verify the entity's actual measures of performance against measure of effectiveness requirements. They always require specific means, sometimes very advanced, more or less easy to implement and expensive. Sometimes, it is difficult to define precisely a measure of effectiveness, because of its subjective and negotiable aspect. Effectiveness testing is more about assessments of actual entity performance than testing. The methodology consists of conducting theoretical studies in order to define the expected measures of effectiveness and the framework of the verifications (worst cases, checkpoints, etc.), then to verify on the entity the real measures of performance, and finally to justify the gap.

Note:

For a non-decomposable elementary entity, there are only functional tests. The architectural aggregation of elementary entities introduces interfaces; these interfaces can generate processing inaccessible by functional tests defined on the basis of requirements such as transformations of input flows into output flows; hence the need for structural testing.

5.3.4.2 DEFINITION OF TEST CASES

Test cases elaboration

Issue with tests completeness

The notion of **performing conditions** is defined in section 5.3.4.1. There are as many test cases to be considered as combinations of inputs, trigger events and valued parameters or constraints; hence the combinatorial explosion of the number of test cases. It is therefore necessary to apply techniques or methods for selecting combinations of performing conditions.

Combinations must be applied:

- Within the frame of definition references for **conformance** testing, looking for possible and impossible combinations, and if all combinations classes have been foreseen.

- Outside the definition references for **defects searching**, or for the characterization of **robustness**, by investigating whether unforeseen final states can exist due to foreseen or unforeseen combinations of the performing conditions (exhaustiveness of the entity's references).

To find unforeseen cases (not defined in the definition references), it is necessary to start from the results and from the final state to raise the intermediate states back to the initial state, by asking the question: if this state and these conditions happen, what are the consequences on the working of the system? While starting from the foreseen inputs (defined), always leads to the expected output states (defined or specified).

Note: To avoid unpleasant surprises in operational use, it is advisable to start by listing all the test cases, then to eliminate some test cases in this list, provide explanation why (impossibility, prohibition, method) and to keep track of these explanations.

Definition of test cases

The test cases are defined for coverage objectives about requirements (functional tests), about architectural characteristics or design properties (structural tests), or for defects searching objectives. The number of defects of an entity is not known in advance; it is therefore not possible to define an objective in terms of the number of defects to be discovered.

In practice, test cases are obtained by defining **generic test cases**, which are then **instantiated** by valuating the performing conditions and the expected results, via the definition of **test data sets** (combinations of the values of the performing conditions and results).

The definition of a generic test case is illustrated in Figure 55.

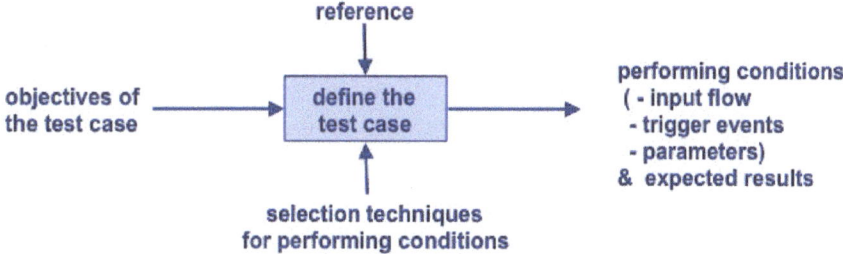

Figure 55 - Necessary elements to define a generic test case

A generic test case is defined by:

- A verification / validation objective; i.e., the execution of one or more functions provided for in the entity's references. A test case should cover only one function of the entity; practically this is not done in order to reduce the number of test cases, or because the reference used is too detailed, or the system level is not appropriate for this test.
- A unique combination of the performing conditions, i.e., input flows, trigger events and parameters - see Note 1.
- Generic expected results (output flows) - see Note 2.

Note 1: The test cases must relate to foreseen performing conditions (valid) AND to unforeseen performing conditions (invalid, i.e., outside the limits specified in the concerned reference).

Note 2: The test cases must relate to what the entity is supposed to do, AND what it is not supposed to do.

An **instantiated test case** is obtained from a generic test case whose performing conditions and outputs are valued. It is therefore defined by:

- a verification / validation objective, i.e., the execution of one or more foreseen functions,
- an initial state characterised by valuation of the performing conditions,
- a final state characterised by valuation of the expected results (output flows).

Valuation is obtained by the elaboration and generation of test data sets; subject discussed below.

Elaboration of test data sets

To exercise the whole of a working domain (nominal - degraded - defined - undefined), it is necessary to have a structured approach, because the natural tendency is to always focus on the same part of the system. This structured approach consists in developing test data sets using different methods or techniques. The notion of the test data set is defined at the beginning of this section 5.3.4; as a reminder, a test data set is a unique combination of the performing conditions, valued, characterized or quantified.

The elaboration of the test data sets is done in two steps:

- **selection** of generic performing conditions, depending on the objectives of the test case, functional testing for conformance verification, or structural testing for defects searching;
- characterization or **valuation** or quantification of these performing conditions to define a particular test, either deterministically (a single value is assigned to each element of the performing conditions) or randomly / probabilistically (several successive values are assigned to each element of the performing conditions according to the adopted probability law) or statistically.

The valuation of the performing conditions makes it possible to determine the values of the expected results.

The elaboration of test data sets is illustrated in Figure 56.

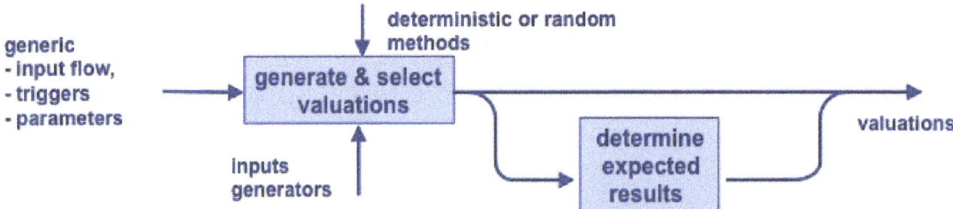

Figure 56 - Elaboration of test data sets

Chapter 5.3. Verification and validation of the system: Practice

Methods or techniques to elaborate test data sets

PARTITIONING ANALYSIS

Principle:

Consists in defining **equivalence classes** on the performing conditions. The functions / actions performed by the entity are **identical** for all elements of the equivalence class.

Therefore, if there is no error in an element of the equivalence class, there will be no error for the other elements of that class.

Identical does not mean same result; because the result depends on the element chosen for the valuation of the performing conditions (input flows, triggers and parameters).

An equivalence class is a subset of which all elements have the same behaviour.

Example:

The entity F calculates the square root of 1/x where x is real.

- Parameter = Real; Input = x; Trigger = no
- There are three equivalence classes: Real zero, Real negative, Real positive. These classes can be quantified in a deterministic way by three different test data sets, using for example the following values: (0.), (-7.), (+5.) [the dot characterizes the Real]

Advantages and difficulties:

- The technique is reliable; it is based on mathematics.
- The difficulty lies in the definition of equivalence classes.

Means:

Either in an empirical way, from the references expressed in natural language, or automatically, if one has a formalized reference expressed in a formal or semi-formal language.

Method:

The method consists of partitioning the input domain into a finite number of equivalence classes, such that one can reasonably assume that a test with a representative element of each class is equivalent to the test of every element of the equivalence class.

The objective of this approach is to minimize the number of test cases to cover all equivalence classes.

The activities of the method are as follows:

- Partition the domain of inputs into a finite number of classes
- Divide each class into two or more subclasses: a class of valid input flows, one or more classes of invalid input flows
- Assign a single element to each class to reduce the number of test cases
- Define test cases to cover valid classes
- Define a particular test case by invalid class, to avoid hiding certain elements

Some criteria for the selection of equivalence classes:

- Interval of values: values in the interval, values outside the interval

Chapter 5.3. Verification and validation of the system: Practice

- Ordered set: each valid element, one or more invalid elements
- Number of elements: no element, exact number of elements, too many elements
- Constraint: respected, not respected

Application to the vending machine:

Reminder of requirements:

1. A vending machine distributes products at 30c and 40c.
2. The vending machine accepts 10c and 20c coins.
3. The vending machine gives the change.

Question: What are the equivalence classes?

Solution: It is not possible to find them directly in the requirements as given above. The method consists of transforming the requirements into a formalized model or scheme. What the developer should normally have done; for example, by using a state-transitions diagram as shown in Figure 57.

But before using the states-transitions diagram (supplied or not by the developer), it must be verified and validated. To verify it, it is necessary to understand how the developer went from the requirements to the diagram. If you do not understand the diagram, you have to talk to the developer, because he made choices when he conceived it. To validate it, one must ask the question of the relevance of the model in relation to the requirements. When the model has been verified and validated (i.e., it is complete and consistent with the requirements), it can be used with confidence.

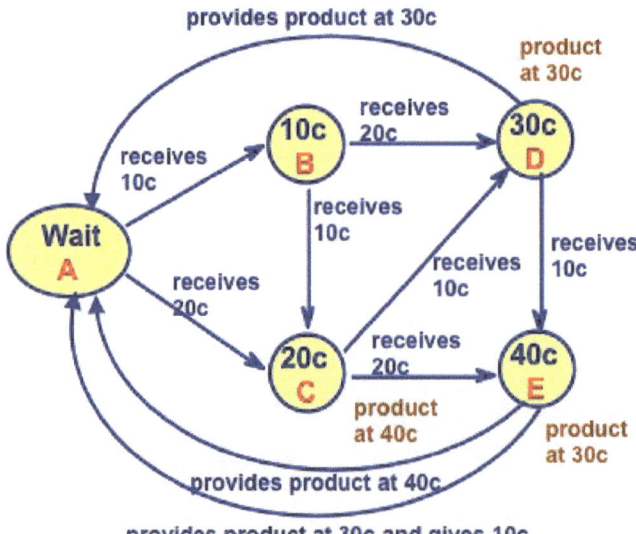

Figure 57 - Vending machine; States-Transitions diagram

There are as many equivalence classes as there are paths between the initial state and the final state of the diagram. The initial state is Wait (A); the final state is Wait (A). It is then sufficient to count all the possible paths between Wait and Wait. To identify the paths, we can use colours or name the states A, B, C, D, E. There are 10 paths, so 10 classes of equivalence. In general, the identification and statement of equivalence classes is guided by the results, NOT by inputs; then we look for the inputs to achieve the results.

In our example, there are 3 results (or output flows):

1) Case of a 30c product obtained with 30c
    a. Class 1: path A, B, D
    b. Class 2: path A, B, C, D
    c. Class 3: path A, C, D
2) Case of a 40c product obtained with 40c
    a. Class 4: path A, C, E
    b. Class 5: path A, B, C, E
    c. Class 6: path A, C, D, E
    d. Class 7: path A, B, D, E
    e. Class 8: path A, B, C, D, E
3) Case of a 30c product obtained with 40c and a return of 10c
    a. Class 9: path A, C, E
    b. Class 10: path A, B, C, E
4) All other cases are cases of error that would need to be detailed on the diagram, or alternatively using other test case development techniques such as, for example, limits analysis.

LIMITS ANALYSIS

Principle and method:

The method of limits analysis consists of defining test data sets for:

- the limit values of each equivalence class of the performing conditions,
- the limit values of the intervals of the performing conditions AND of the observable elements.

It consists of checking the actions performed by the entity under test (result provided) for each equal, immediately lower and immediately higher value of each input, and/or each equivalence class or the interval between the input values.

Limits analysis can also be done on all observable elements, selecting performing conditions that generate results (outputs), or checkpoints at the lower and upper limits.

> Limits form the natural host of most defects!

Here are some test cases selection criteria:

- For intervals of values:
    ♦ Each lower and upper limit
    ♦ Each limit ± delta

Chapter 5.3. Verification and validation of the system: Practice

- For sets of ordered elements:
  - The first and second element
  - Before last and last element
- For a number of possible elements:
  - Minimum number
  - Maximum number
  - 0 and 1 if it is possible

Example:

Let the entity F, which for x integer, performs the following calculation:
- If x belongs to the interval [-10 , 10] → x
- If x belongs to the interval]-100 , -10[or]+10 , 100[→ 2x
- If other intervals or values → 0

The variable x is an integer, the difference between 2 integers in this case (delta) = 1.

The test data sets to be defined relate to limit values and around the interval limits, namely:
- -100, -101, -99, -10, -11, -9
- 10, 11, 9, 100, 101, 99

Advantages and difficulties:

- Errors at the limits reveal real logic defects that can occur during normal use of the entity.
- It may be difficult to define the limits of output values or checkpoints; at least the inputs (performing conditions) necessary for the entity to provide outputs or results at the limits.
- Errors at the limits can compensate for each other and, in this case, they do not appear. It is therefore necessary to verify each limit value independently.

Application to the vending machine:

Equivalence class tests are completed with limits tests; that is to say what happens at the junction between the equivalence classes.

This time we are interested in the input flows, i.e., to all the possible coins. Apart from 10c and 20c, the vending machine should reject all other coins: 1c, 2c, 5c, 50c, 1 Euro, 2 Euros.

STATE-TRANSITION MATRIX

Principle and method:

The state-transition matrix allows exploring the combinations of the performing conditions. The method consists of representing the references, or a part of them, in the form of a States-Transitions diagram. On each transition, the events (triggers) that lead to a change of state and the triggered actions are expressed. This diagram is then translated into a matrix or table by arranging the states on the columns of the table and the events or triggers on the lines. Each cell of the table is at the

Chapter 5.3. Verification and validation of the system: Practice

junction of a present state (column) and the arrival of an event (line); it contains the next state and the action triggered by the transition.

Example:

Look at the States-Transitions diagram in Figure 58.

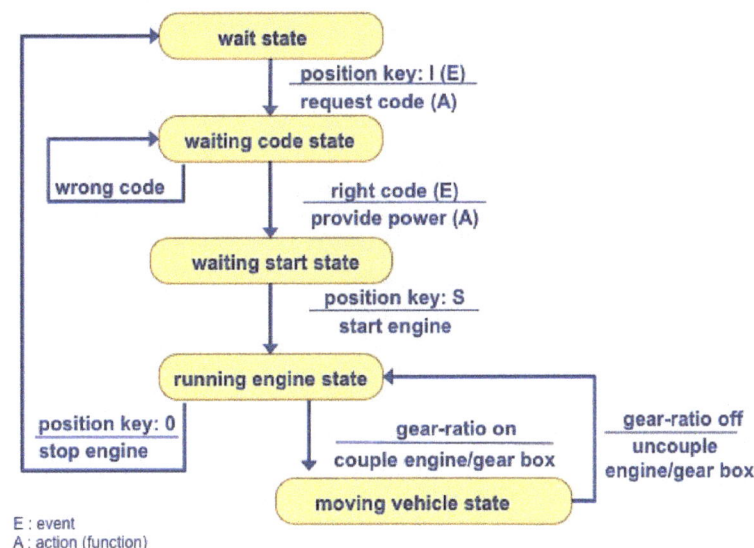

E : event
A : action (function)

Figure 58 - Example of a States-Transitions diagram

And the corresponding matrix in Figure 59.

events \ states	wait	waiting code	waiting start	running engine	moving vehicle
position key: I	request code / waiting code	?	?	?	?
wrong code	?	do nothing / same state	?	?	?
right code	?	provide power / waiting start	?	?	?
position key: S	?	?	start engine / running engine	?	?
position key: 0	?	?	?	stop engine / wait	?
gear-ratio on	?	?	?	couple engine/gear box / moving vehicle	?
gear-ratio off	?	?	?	?	uncouple engine/gear box / running engine

Figure 59 - Example of a States-Transitions matrix

173

Chapter 5.3. Verification and validation of the system: Practice

Advantages:

- States-Transitions diagrams are developed from the operational scenarios. Normally developers have modelled the operational scenarios. It is then possible to translate them into States-Transitions diagrams and then into States-Transitions matrices.
- The States-Transitions matrix allows crossing the different states with all the possible triggers.
- It allows the identification of all combinations of unspecified performing conditions (all "?" in the figure) that could cause problems (possible case, impossible case, dangerous, forbidden, etc.).
- It provides combinations of performing conditions for defects searching tests.

TRANSACTIONAL ANALYSIS

Principle and method:

Transactional analysis is used to define test data sets that sensitize transactional flows; i.e., the sequences of operations that allow the more or less complex exchanges of matter, energy and/or information flows between two or more actors / elements internal and/or external to the system.

The method consists of representing the reference, or part of it, in the form of a transactional flow diagram, in which:

- a transaction is:
  - a functional unit from the user of the system viewpoint,
  - a sequence of operations or tasks (executed by the system, actors or elements external to the system) from the system viewpoint,
- a stimulus triggers the transaction,
- a transaction flow is a representation of the path taken by the transaction via the execution of a sequence of functions / actions by elements exchanging matter, energy and/or information flows,
- the arcs in the diagram link actions or tasks,
- the nodes in the diagram are junction and separation points between the transactional flows.

Each path in the diagram between the starting point and the end point of the transaction describes a test case.

Example:

Figure 60 represents the transaction between a Cash Machine and a user.

Chapter 5.3. Verification and validation of the system: Practice

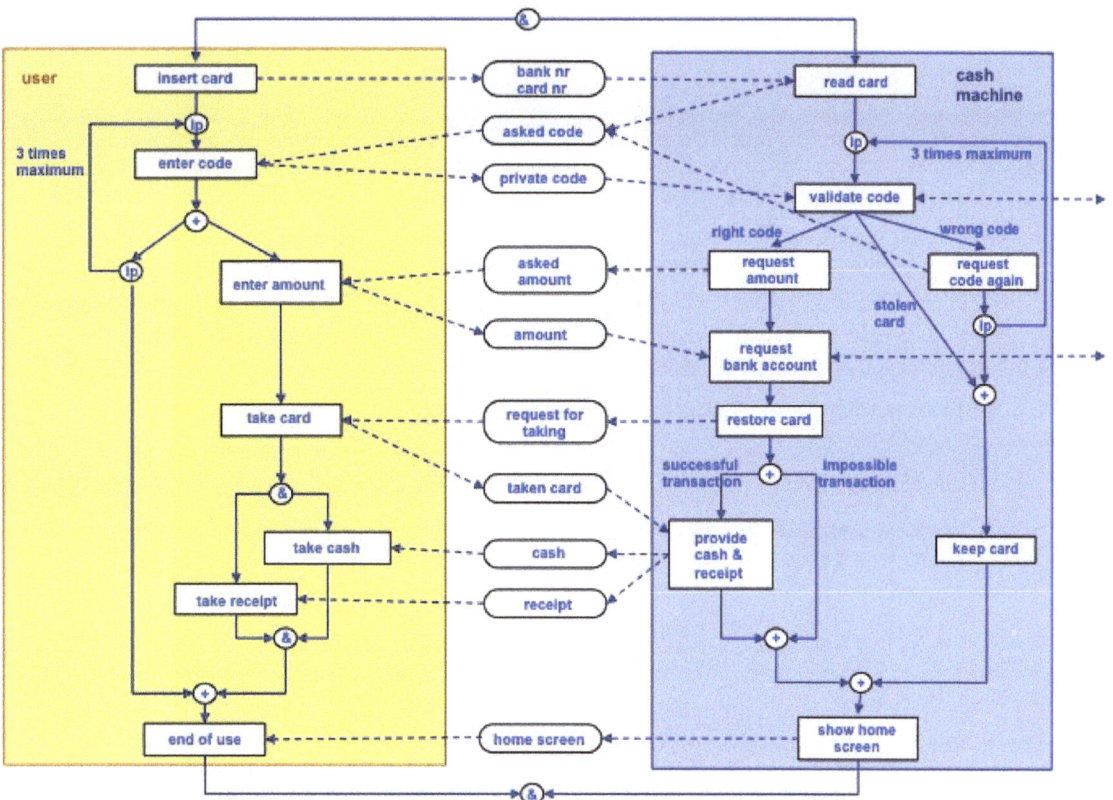

Figure 60 - Example of transactional flows - Cash Machine

In this figure, the diagram uses the principles of eFFBD (enhanced Functional Flow Block Diagram). The symbol & is a logical AND, + is a logical OR, lp is an iteration / loop. The two dotted arrows to the right of the diagram represent information exchanges with the credit card consortium (external element to the system).

Advantages:

- Transactional analysis makes it possible to identify and define the flows exchanged, i.e., the interfaces between the concerned elements.
- It allows validating the logic of the transactions using representation models, and consequently the operational scenarios seen by the users of the system, through the logical solution envisaged during the definition and the implementation of the architecture and the design.
- It is also a technique to represent operational scenarios including human actors (operator role - in the system, user role - external to the system); including human error cases.

Chapter 5.3. Verification and validation of the system: Practice

Random generation of the valuation of test data sets

Principle and method:

We have previously said that the **valuation** or quantification of the performing conditions to define a particular test was done either deterministically (only one value is assigned to each element of the performing conditions) or randomly / probabilistically (several successive values are assigned to each element of the performing conditions according to the chosen probability law) or statistically.

In deterministic valuation, the choice of values is often guided (sometimes unconsciously) by the solution as conceived and implemented. The test data sets are then statistically biased and do not reveal certain defects with respect to the expected behaviour (requirements).

The random generation of the valuation of the performing conditions is performed according to a probability law; it can be three types of law:

- **uniform random** law, that is, a uniform law on the domain of inputs,
- **operational random** law, that is, a law representative of the predictive distribution of operational conditions,
- **structural statistical** law, i.e., making the execution of certain internal structures (e.g., the two branches of a decision) almost equally likely.

The probability law applies to the input domain. It can be defined by an analytical technique (study of the structures of the solution) or by an empirical technique by successive refinements of the initial law.

The probability law and the duration of the test must be rigorously chosen according to the objectives of the test and the associated criteria used. The characteristics or values are drawn randomly: it is necessary to carry out several draws for the same test case before being able to conclude on the case. The number of draws required, and therefore the duration of the test, is very important.

Random generation requires a methodical approach, because the choice of a uniform distribution is not a miracle solution or a guarantee of efficiency in the detection of defects.

Advantages:

- ♦ The generation of the valuation of the performing conditions is automatable.
- ♦ Automatically generated characteristics or values are more objective than with deterministic valuation.
- ♦ Reliability measurements are possible due to the number of test cases generated.

Difficulties:

- ♦ Preliminary definition of probability laws that reflect operational conditions or allow for equiprobable executions.
- ♦ The workload required to characterize or value the expected results (outputs) is rarely automatable.
- ♦ The duration of the tests is generally long as each generic test case is run with several series of characteristics / profiles / values.
- ♦ When the input domain has been poorly circumscribed, the random generation of test data sets may not permit to execute specific behaviours (special cases, error cases, exception processing).

Chapter 5.3. Verification and validation of the system: Practice

5.3.4.3 IMPLEMENTATION OF TESTS

Overview of the implementation of tests

The implementation of tests includes activities relating to the product / service / organization, activities relating to testing facilities and joint activities.

Figure 61 illustrates these activities and the main flow.

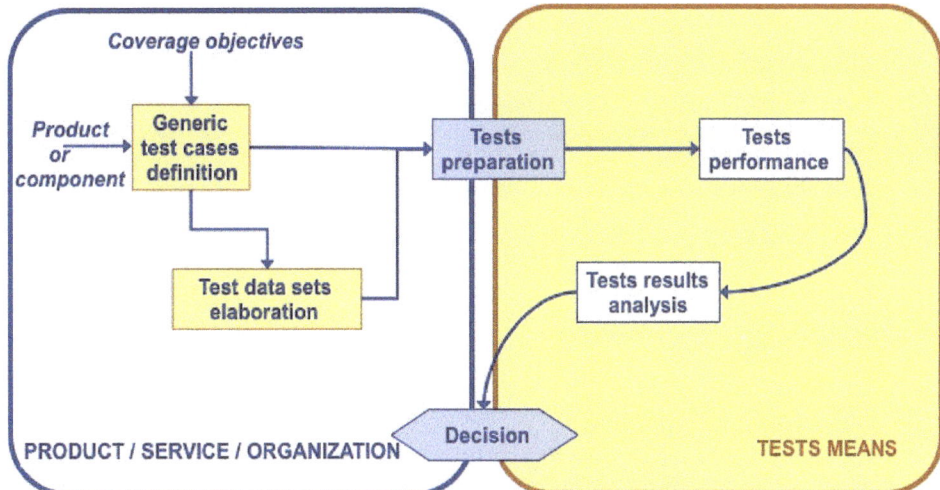

Figure 61 - Activities of tests implementation

- **Activities related to the product / service / organization** (or their components) to be tested include:
    - the definition of generic test cases according to the coverage objectives: coverage of the requirements, coverage of architectural characteristics and design properties, coverage of implementation characteristics;
    - the preparation of the flows necessary for the product / service / organization to be executed, that is, the valuation of the performing conditions in the form of test data sets: valuation of input flows, triggers and parameters.
- **Activities related to testing facilities** include:
    - the execution of the test procedures that drive the stimulation of the product / service / organization;
    - the recording of test results and their analysis by test means.
- **Joint activities** with the product / service / organization and testing facilities include:
    - the preparation of tests: definition and implementation of test procedures, files used by test means containing data and instructions described with the appropriate languages and formats;

- the decision-making on:
  - the conformance of the obtained results with the expected results,
  - the assessment of the requirements coverage and/or architectural characteristics and design properties coverage,
  - the conformance and adequacy of test means and test procedures.

Focus on testing facilities

The means dedicated to testing may be very important; so, they have direct impacts on the decision about whether or not to carry out a particular type of test.

Specific test means, or tools, correspond to each activity previously listed.

The tools for defining generic test cases depend on the type of test to be performed. There are many tools for testing technological components and for integration testing. Tools for final product / service / organization validation tests are rare. Techniques for selecting the performing conditions help to define relevant elements, for example the distribution in equivalence classes.

For the development of test data sets, it is possible to use generators or simulators of varying complexity; for example: image processing, data transmissions, biological or physic-chemical processes. The difficulties in producing the input generators relate to the approximation of the reality (thus the validity of the input elements and the representativeness of the tests in relation to the actual expected operation), the quality of their development and their tuning or tweaking.

The tools to assist in the preparation of the tests mainly concern the development of test procedures, which use languages (often specific) compatible with test benches and related data files.

The test execution tools are named test benches. These test benches can be composed of real equipment, simulated equipment (the behaviour is simulated: to such input stimuli corresponds with such output), environmental simulators, test control devices, on-line or off-line investigation tools.

The decision-making tools are analysis and archiving tools that allow the storage of recorded results on-line and to analyse them off-line.

Remark:

It is very common that the target system (product, operational service) serves to validate the simulators, and vice versa, in parallel; so-called back-to-back validation. This practice is quite normal, as simulators and target systems are not conceptualized, developed or produced using the same methods and/or the same viewpoints.

Tests definition

The activities for the definition of generic test cases and for the development of test data sets are described earlier in section 5.3.4.2. The reader can refer to it. The information contained in a test case definition sheet is as follows:

- identification of the test case, title
- objectives of the test
- initial state of performing conditions (input flows, parameters)
- external actions to carry out (stimuli)

- final state (expected results)
- intermediate states (internal flows for structural tests only)

Preparation of tests

This activity consists of answering the question: what are the necessary elements to carry out one or several test cases? It includes tasks to develop test procedures from defined test cases, preparation of the target product, and preparation of test means and tools.

A test case (sometimes referred to as a test specification) describes the objectives of the test, the functions / actions to be performed and the constraints to be met - see above.

A test procedure usually groups several test cases (test sequence). It describes the operations to be carried out by a human operator or in automatic mode, the operational or organizational conditions required to run the concerned tests - see below.

Development of test procedures

The development of the test procedures consists of defining and ordering the tasks to be carried out in order to chain the execution of the test cases - see synthetic presentation in Figure 62.

The development activities for a test procedure are described below:

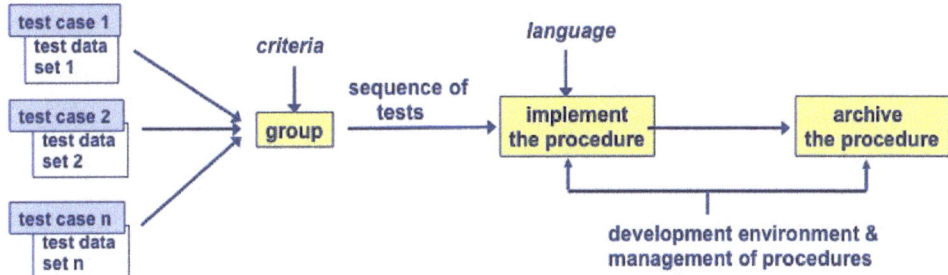

Figure 62 - Development of a test procedure

- **Group test cases** according to criteria. Groups of test cases are mainly used for integration and final validation testing (on industrial site or on operational site). The grouping criteria mainly concern the performing conditions, for example:
  - the initial performing conditions (or initial state) of case i = final state of case i-1 (or results),
  - actions of test cases triggered by the same event,
  - same performing conditions for parallel actions,
  - sequence of non-separable sub-functions at the concerned test level.
- **Constitute a sequence of tests** - see Figure 63. An operational scenario can very well be used as a sequence of tests. A test sequence is the result of a grouping of test cases and associated test data sets (performing conditions) in which:
  - the parameters of the different test cases are identical or exclusive and are not contradictory,

Chapter 5.3. Verification and validation of the system: Practice

- ♦ the inputs of the different test cases are compatible,

- ♦ the stimuli or events of each test case can be linked in time,

- ♦ the results of the actions generated by the test case i are inputs for the test case i + 1.

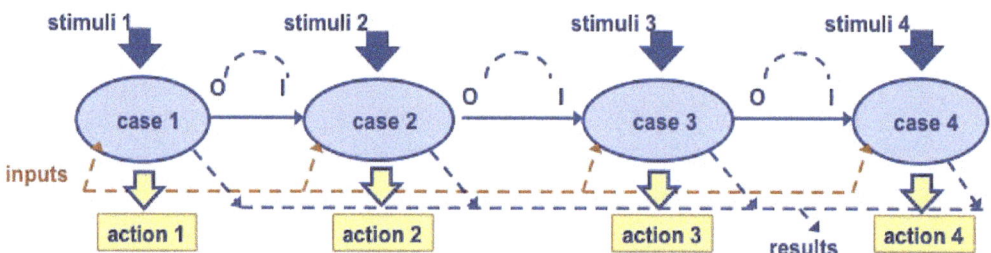

Figure 63 - Sequence of tests

- **Implement the sequence in a procedure** using the specific language of the test means or tool used. The language can consist of manual commands, or automated commands via a computer tool. This computer tool includes software with specific commands for debugging: start, stop, run, pause, restart, step by step run, data to record, trace, box message, etc.

The information contained in a test procedure sheet is as follows:

- ♦ list of included test cases

- ♦ description of the test environment: configuration of necessary hardware and software and their initialization

- ♦ definition of the initial state of performing conditions (input flows, parameters)

- ♦ ordered description of commands (stimuli) to be applied over time, step by step (manual commands or automatic simulator or tool-specific commands)

- ♦ checkpoints and results to be recorded

- ♦ checkpoints to be checked during the execution of the procedure

- **Verify and validate the procedure**. Preferably the test teams write and verify the test procedure; the engineering teams validate it against the system's definition references.

- **Archive the procedure** for immediate or delayed execution. Test procedures are managed using development and archiving tools to enable coding, generation of executable, archiving, and versioning of automated procedures and associated files.

Remark:

For the test of elementary components, the test procedure is very often reduced and rarely formalized, as it requires few external commands; the workload would be too important for uncertain profitability. Other means may be used.

Preparation of the target product

The target product is the operational product or operational service or operational organization. The term **product** is used here to cover the three major types of systems.

At the level of the final validation of the product, whether on an industrial site or on an operational site, the target product must be in the condition in which it will be delivered, except the corrections to defects, i.e., disconnected from any non-contractual instrumentation. Whereas at the level of integration or subsets testing or tests of elementary components, it is possible, even desirable, to instrument the product or parts of it, for the purpose of observing internal behaviours via the introduction of checkpoints.

When assembling subsets of the product or integration aggregates, the implementation of the tests requires the addition of harnesses, plugs, simulators, stub modules or driver / launcher modules - see illustration Figure 64. Observation of the internal behaviour of the target product requires more or less sophisticated external means, such as probes, sensors, spyware on input-output buses, addressing buses, plotters, specific analysers, step-by-step monitor, etc. The instrumentation is mostly specific to the application or the product, and therefore not generalizable. It is an important part of the overall testing workload. The costs associated with the instrumentation dedicated to testing must be provided as soon as possible.

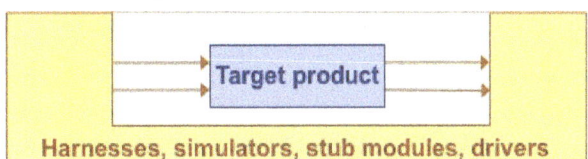

Figure 64 - Preparation of the target product for testing

A test driver / launcher is strongly linked to the part of the tested product from which it consumes resources. This may adversely affect the actual performance and effectiveness of the product or may reveal errors related to the configuration surrounding the part of the target product being tested. The analysis of the results must, therefore, take into account the test configuration before concluding that the target product is faulty or defective.

Preparation of test means

The means used for carrying out the tests are named *test benches*. A test bench typically includes:

- a control test unit for operator access, and for conducting test case sequences; the control test unit is linked to the instrumentation and allows the behaviour of the target product during operation to be observed via the checkpoints;
- environmental simulators, harnesses, input flows generators;
- means for recording collected measurements and test results;
- safety equipment for operators;
- access to energy resources.

The tasks for preparing the test means consist of connecting all the means listed above to the target product, during their configuration and initialization: hardware, software, input flows generators, means for collecting measurements at checkpoints and test results.

The test procedure sheets contain the information necessary to prepare the test means.

Performance of tests

The performance of the tests is in fact the execution of the test procedures, whether these are manual or automated. This execution is organized in planned **sessions** in the test plan of the concerned level; this test plan (for components, integration, or final validation) is usually a part or sub-plan of the verification and validation plan or the integration plan.

Each session is composed of - see Figure 65:

- a Tests Preparation Review (TPR),
- the execution of test procedures,
- a Tests Results Review (TRR).

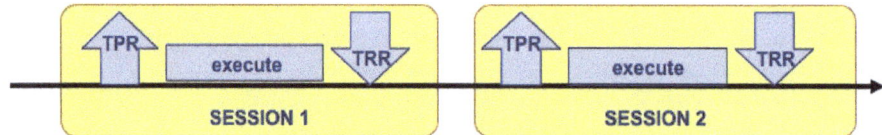

Figure 65 - Run of tests sessions

The **Tests Preparation Review** (TPR) consists of verifying:

- the existence of the test cases list
- the existence of test data sets
- the existence of corresponding test procedures
- the status or configuration of the tested product: identification / version of the product, of its components and the associated documentation
- the status of Deviations, Non-Conformance Reports and Change Requests
- the status or configuration and availability of test means: identification / version of hardware, software, associated documentation
- the availability of human resources (operators)

The execution of test procedures consists of:

- executing the test procedures one by one in the order defined in the test plan;
- deciding on the end of the session on criteria: end of normal execution of the procedures, identification of a blocking anomaly of the target product or test means working;
- producing a complete test report and a summary (number of test cases and procedures performed with or without anomalies, number remaining to be run in case of shutdown);
- drawing up any anomalies sheets.

Notes:

An anomaly is an abnormal behaviour of the target product and/or test means when a non-conforming result is produced, or a result of the target product is unavailable, or a test device malfunctions.

Chapter 5.3. Verification and validation of the system: Practice

The modification of the target product is prohibited during the test session, unless the anomaly leads to a blocking defect for the subsequent tests. Corrections or repairs are required at the end of the session. The same session can be re-executed on a new version of the target product.

The **Tests Results Review** (TRR) consists of:

- analysing the tests reports,
- analysing the anomalies sheets,
- producing a tests summary for decision making.

Note:

The Tests Results Review defines, in particular, the tests that must be re-executed following modification, correction or repair.

Analysis of tests results

Archiving tests results

Test results are recorded when the test procedures are executed; they are generally analysed off-line and recorded in a test analysis report. Archiving is carried out for validation tests of the system (product, service, organization), more rarely for elementary component tests or integration tests.

The information contained in a test result analysis report is as follows:

- identification of the executed test procedure, title
- configuration of the tested product (version), person responsible for the test
- test cases covered
- test environment (real configuration of used tests means)
- list of execution files (files for inputs, results, execution traces)
- obtained results
- encountered anomalies (reference of anomaly or non-conformance sheets)
- conclusions (OK or not OK)

The archiving means are essentially mass memories used by specialized workstations, which also include tools for helping to analyse the results.

The archiving of the test results not only allows comparison of similar tests or the same tests carried out under similar conditions, but also to compare the results obtained on test facilities with the results obtained on the target product in operation to investigate causes of anomalies found on the target product. Finally, it allows the verification of the non-regression of the product during the maintenance phase - see section 6.3.4.

Analysis of tests results

The basic issue is to decide on the success of the test; i.e., to be able to answer the question: how to determine if the obtained result conforms with the expected result? This is called the oracle problem.

If an anomaly is found, it is recommended to thoroughly analyse this anomaly to identify its cause. It may come from the target product, but also from the other elements used in the test: test means, simulators, test procedure and test data sets - see Figure 66.

The origin of the anomaly must be suspected from the left to the right of the figure: first suspect the test procedure, then the implementation of the means, etc.; the suspicion of the target product is the last step to consider.

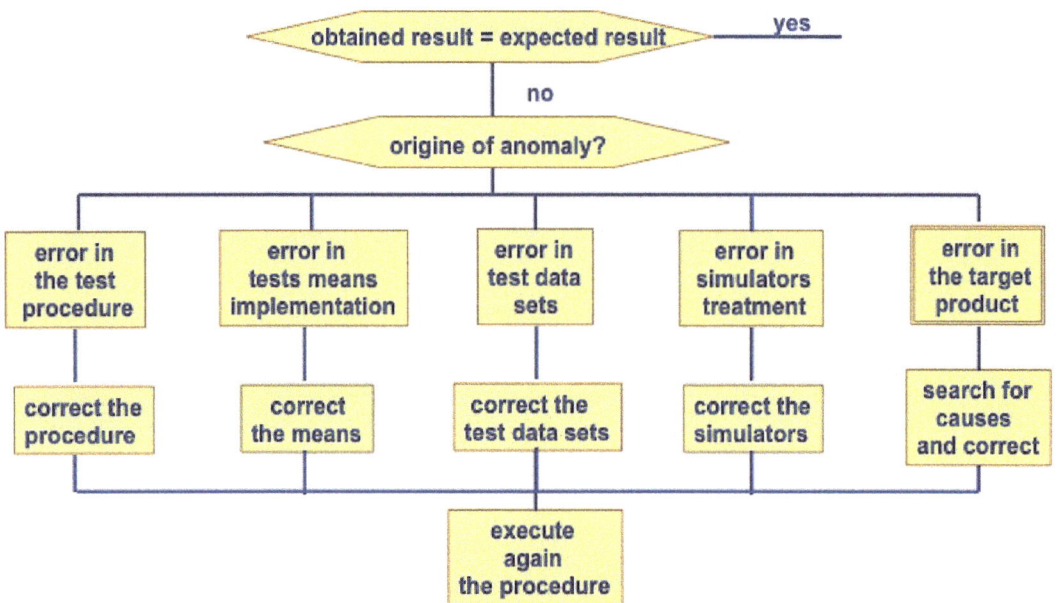

Figure 66 - Analysis of the anomaly

In the case of functional tests and deterministic test data sets, there is no ambiguity for the conformance of the Boolean results or over a range of expected values; conformance can even be determined automatically. Complex behaviours that can be analysed in the form of variables evolution (often represented by curves) are more difficult to analyse.

In the case of structural tests and random test data sets, it is difficult to predict the expected results. In the case of critical systems, one method consists of developing prototypes or variants of the concerned component, starting from the same requirements, but developed with different models and means; then compare the results obtained by the different variants (back-to-back testing). The subsidiary question is: what is the right variant? It is customary to retain the variant that produces the least errors during the tests. Formal specifications seem to be the best technique for releasing reliable oracles.

End of testing

The decision to stop testing should not be taken because there is no longer the time or budget. This issue is an integral part of the development of the verification and validation strategy - see section 5.1.5. At each end of test sessions, the question must be asked. The answer is given objectively according to the coverage rates reached at that particular time of the tests campaigns. Figure 67 summarizes the questioning and the evaluations to be carried out, in order to decide whether to continue or stop the tests.

Chapter 5.3. Verification and validation of the system: Practice

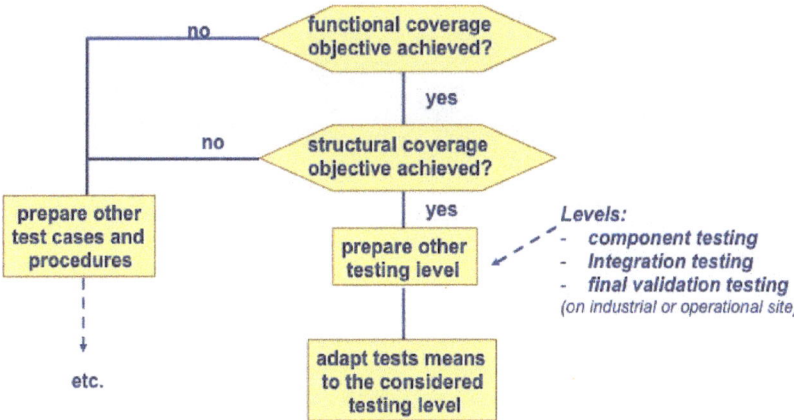

Figure 67 - Decision to stop or continue testing

The test means and environment evolve with the progress of the target product testing:

- partial means and environment at components testing level,
- means and environment for verifying functional chains on aggregates with automated procedures at integration testing level,
- almost real environment (if it is possible) at final validation level on an industrial site,
- real environment at final validation on the operational site.

Actors for testing

Several types of skills are required to implement the tests. The activities related to testing should be distributed among the actors as follows:

- The definition of the tests is carried out by the persons in charge of the establishment of the three or four major reference documents or files, namely system requirements, architecture description, design description, and implementation description.
- Test preparation, test performance and analysis of results are carried out by test specialists: technological components tests specialists, integration tests specialists, and operational tests specialists.
- Analysis of test results and subsequent decisions is carried out jointly by specialists in the field of testing and by those responsible for the concerned references.

5.3.4.4 SYNTHESIS AND EFFECTIVENESS OF TESTS

Table 10 presents, in a synthetic way, the typology of the tests according to the target objectives.

A defect-searching test is effective if it finds many defects with undesired consequences.

A conformance test is effective if it demonstrates compliance with the references at the lowest cost and with maximum confidence.

	Conformance (test defined against definition references)	**Defects searching / robustness** (test defined outside definition references)
Functional tests **black box**	1) Combinations of the performing conditions that lead to nominal expected states	3) All other combinations of performing conditions that are tested element *behaviour-oriented*
Structural tests **grey box**	2) Combinations of the performing conditions that lead to non-nominal expected states	4b) Linked to design decisions, oriented towards *not accessible, unrealized or not foreseen parts*
Structural tests **glass box**	4) None - except required dependability	5) Linked to implementation decisions oriented towards *undesired dysfunction*

Table 10 - Tests mapping

The preferred order to define and perform tests is given by the number in each cell of Table 10.

5.3.5 Organization of verification and validation activities

Psychology and culture

In general, developers see verification and validation as merely an ended formality that will only confirm with technical choices; only a few minor changes would be made! As a consequence, development staff has little interest in verification and validation processes, and does not seem concerned with the *testability* of products.

As the reader of this book has discovered, the activity of verification and validation is something else. It is as necessary as the system definition activity. In fact, the two activities must collaborate from the very beginning of development, to delivery. The verification and validation activity yields fantastic feedback on the definition and the implementation, without which no quality is guaranteed, and no improvement is possible.

Nevertheless, when the activity of the developer is fundamentally positive and creative, the activity of the verifier is rather negative and destructive: he or she focuses on what does not work by looking for defects, non-conformances. It is therefore necessary to have a somewhat peculiar mind that seeks to see what is not working properly, but with the very noble goal of the satisfaction of the acquirer, and obviously of the developers, since he or she participates in the improvement of the development.

The profession of verification and validation is subject to little formalization and little promotion, yet which is very enriching, as it requires special technical and human qualities.

Organization and responsibility

A complete verification - validation team is typically composed of four roles, under the responsibility of one person, the verification - validation manager. The roles and competencies are as follows:

- Reviews specialist
- Integration and final validation testing specialist
- Target product / service / organization specialist
- Quality monitoring specialist

The person responsible for verification - validation could be:

- The project manager: this is not a good option, because in the project there are two different views, apparently contradictory because of their psychological orientation, but which together turn the wheel. A single head for different orientations is very difficult to live.
- A deputy to the project manager, a verification - validation expert, who only performs this function: it is an excellent option, as long as they collaborate from the beginning of the project; everything will go faster for an optimal quality related to the constraints of the project.
- The quality engineer of the project: this is an option not to be neglected if he is a real assistant to the project manager, in spirit at least.

5.3.6 Recommendations and frequently asked questions

It is impossible to verify everything in any system, because of cost and time constraints, availability of skills and some timely test infrastructures, etc.; also because of environmental constraints, products or organizations existing in the context of use. Consequently, deadlocks are inevitable, which de facto creates potential risks: too many residual defects during delivery or commissioning, or confidence of the acquirer not achieved.

> **The only solution of not being able to check everything is to define verification actions and validation actions relevant to the detection of defects, and to obtain confidence.**

The main pitfalls identified in many development projects and leading to serious inconvenience are listed below, with a recommendation for each:

- ♦ Save time by not processing verifications throughout the development!
    - The consequence is a waste of time at the end of development. Decision to be banned definitively.
- ♦ Only carry out validation tests at the end of the development!
    - Set up relevant verification and validation techniques during system engineering (definition phase).
- ♦ Stop verification and validation actions on the expiry of budgets or deadlines!
    - Define objective criteria, such as coverage rates, with risk assessment if verification and validation actions are excluded.
- ♦ Perform unnecessary verification and validation actions, such as performing the same tests, due to the verification techniques and/or integration techniques used!
    - Define criteria to select relevant verification and validation actions.
- ♦ Late start of verification and validation activities in the project!
    - Organize verification and validation activities as support (feedbacks) to system definition and integration.

Chapter 5.3. Verification and validation of the system: Practice

FAQ

What do the terms "verify a system requirement" and "verify a system" mean?

"Verify a system requirement" means that its textual expression has correctly used syntactic and grammatical rules (subject, verb and complement), and that it has the following characteristics:

1) necessity - the requirement is necessary for the concerned system in its context of use,
2) unambiguous - there are not several possible interpretations,
3) coherent - the set of terms used form an expression having a meaning,
4) complete - all the elements necessary for its understanding are in the formulation,
5) feasible - the expression describes a feasible characteristic,
6) traceable - it is possible to find its origin,
7) verifiable - there is an appropriate technique to demonstrate compliance with the concerned product.

"Verify a system" means that the characteristics or properties of the product, service or organization to which the system relates must conform with the expected architectural characteristics and design properties.

What do the expressions "validate a stakeholder requirement", "validate a system requirement" and "validate a system" mean?

"Validating a stakeholder requirement" means finding that its textual expression, as meaningful content, is justified and relevant to the expectations of the concerned stakeholders, complete and expressed in the language of the customer or user.

"Validating a system requirement" means finding that its textual expression, as content meaning, correctly and/or accurately translates one or more stakeholder requirements into the language of the supplier-designer.

"Validate a system" means that the characteristics or properties performed on the product, service or organization, to which the system refers, conform with system requirements and/or stakeholder requirements.

What is the difference between "verification action" and "verification activity"?

A *verification action* concerns the object (system, component, product, service, organization, etc.) submitted to verification in relation to its definition reference, using a verification technique or method (e.g., inspection, test). See section 5.1.2.

A *verification activity* involves the verification process, which consists of activities; each activity can be composed of more basic tasks. See section 5.2.1.3.

Should a test case be defined to justify compliance with each requirement?

To justify compliance with a requirement, the first idea is often to perform a test. But a test requires that the system exists (at least in the form of a prototype), to implement a suitable tool (often expensive and which must be validated), to construct a representative test. Other techniques (inspection, studies, etc.) can be more effective and easier to implement. For example, measuring reliability using working tests may need to wait until the system is at the end of its life! This is why assessment techniques are used for reliability estimate (see Volume 5 of this series of books).

How to reduce the number of test procedures to be performed?

A good practice consists of grouping the test cases together, taking as a common thread the operational scenarios established by the client and/or the scenarios established by the developer / architect; and then to define test cases to cover specific situations and properties.

More generally, the sorting of the Justification Matrix according to the verification - validation techniques, and by phase / development activity makes it possible to select the verification or validation actions according to the phase or milestone for execution. This makes it possible to organize and optimize the verification - validation strategy and plan.

Is a document signed by the hierarchy validated?

The signature of a document by the hierarchy is not the justification for its validation. The signature does not attest that the document was the subject of a validation action. It only provides authorization for the distribution of the document outside the authors' circle and their hierarchy (functional or operational).

What is an "experiment plan"?

An experiment plan is an orderly sequence of tests performed during an experiment. Each test allows the acquisition of new knowledge by acting on one or more input parameters to obtain results validating a model.

Experiment plans can be used to evaluate or optimize certain system properties.

In the area of verification and validation, experiment plans can be used to optimize a verification - validation strategy. They also make it possible to optimize the choice of tests by relying on feedback from similar products or services.

What is acceptance?

Acceptance is an activity carried out before the delivery of the system (product, service, organization) in order to decide the change of ownership of the system from the supplier to the acquirer. A set of operational validation actions is often performed, and/or a review of the validation results is systematically carried out.

Chapter 5.3. Verification and validation of the system: Practice

What is certification?

One distinguishes the company or organization certification, and product certification. In all cases the certification is carried out and pronounced by an external and independent accredited body. Company certification is the examination of practices, processes against a regulation or a standard (example ISO 9000, ISO 14000).

Product certification is the examination of the characteristics of a product, as well as the practices, methods and/or processes applied for its development and/or manufacture, in relation to standard(s) and/or regulation. Development reviews, results of verification and validation actions form the basis for product certification.

Product certification is a written warranty that the product may perform the functions assigned to it and that it has been developed and manufactured or produced in accordance with applicable and accepted standards or regulations.

What is qualification?

Qualification is an activity. This term is not standardized across industries, and qualification activities may be different. Some industries use this term for the final validation activity on an operational site. For other industries, it consists of qualifying or accepting the system under operational conditions, before commissioning / operation, or after a certain period of operation demonstrating that the system conforms with its expectations (implemented system characteristics or properties, including margins, satisfy applicable system requirements and/or stakeholder requirements).

The qualification of the system requires that all verification and validation actions have been carried out successfully beforehand. The qualification is concluded during an acceptance review.

This term is also used for the qualification of personnel in certain professions.

6 INTEGRATION OF SYSTEM COMPONENTS

This chapter presents the necessary means to carry out the integration of the components that compose the system.

This chapter presents:

- The integration concepts and approaches - Chapter 6.1 which includes:
  - Definitions and concepts - section 6.1.1
  - Integration approaches - section 6.1.2
  - Integration strategy - section 6.1.3
  - Relationships of integration with verification and validation - section 6.1.4
- The process approach - Chapter 6.2, which includes:
  - The location of the process in the development cycle - section 6.2.1
  - The purpose, inputs and outputs of the process - section 6.2.2
  - The description of the activities of the process - section 6.2.3
  - The ontology elements - section 6.2.4
  - The artefacts generated by the process - section 6.2.5
- The practice - Chapter 6.3, which includes:
  - Heterogeneity of the components and complexity of the interfaces - section 6.3.1
  - Notion of integration enabling system - section 6.3.2
  - Interfacing with other development processes - section 6.3.3
  - Searching for defects and clearing defects - section 6.3.4
  - Recommendations and FAQ - section 6.3.5

6.1 Concepts and approaches for integration

6.1.1 Definitions and concepts

> **Notion of integration**
>
> The integration of the system is the set of activities that combine several components of the system and activate the interfaces, to form a realized system (product, service, or organization), which allows interoperation between the components of the system and other systems, in order to satisfy system requirements, architectural characteristics and design properties of the system.
> (From ISO/IEC TS 24748-6 [ISO 24748-6]).

The execution of these activities produces something **integrated**; something that becomes a **whole**. Globally, a system, whether conceived or realized, is more than the sum of its parts: it forms a whole endowed with emergent properties.

Examples of integrated things:

- The engine is integrated into the ground connection (chassis, suspension) and then to the passenger compartment to form the vehicle.
- Integrated circuit: a small piece of semi conductive material that contains interconnected electronic elements.
- Integrated team: a group of people with complementary skills and expertise, who are committed to delivering specified work products or defined results in timely collaboration.

Integration is more than the simple assembly of components. Assembly consists only of the physical union and connection of the components.

The integration of the system consists of receiving the various implemented components that compose the system (in fact the product, service or organization), assembling them physically or concretely, performing verification actions on the system during the assembly and at the end of assembly. Integration includes verification of system functions. The scope of integration is schematized in Figure 68.

The purpose of integration is to obtain a physical product, service or organization that conforms to the description of the architecture in order to validate experimentally the solution against the requirements.

The final validation activity can be separated from the integration activities, or not, depending on whether or not the integration of the system into its operational environment or into a higher system level is considered. This point is discussed later.

> The concept of **wholeness** is the key to the integration engineering of the system.

Chapter 6.1. Integration: Concepts and approaches

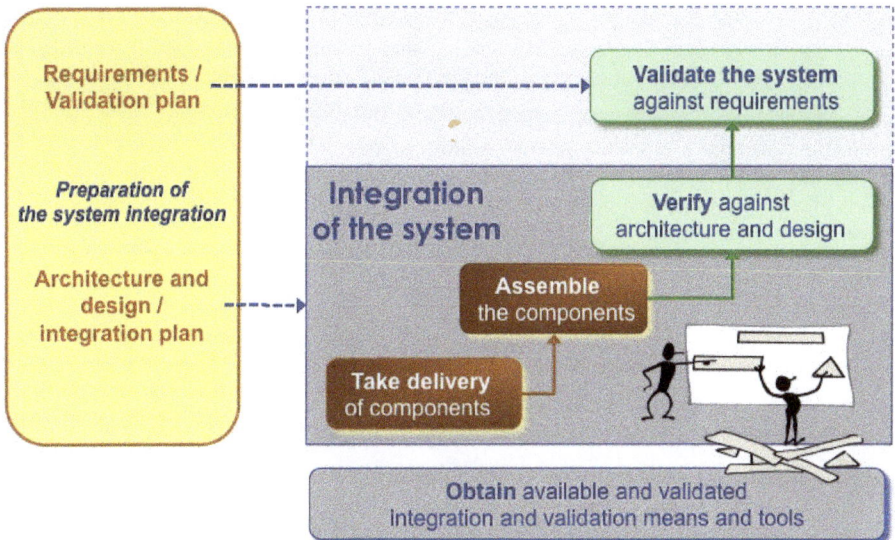

Figure 68 - Scope of system integration

Notes:

Separating integration activities from verification and validation activities makes no sense from a practical point of view as they are linked. Nevertheless, this separation is an excellent intellectual exercise, which makes it possible to clarify the concepts used and their relations, and also to clarify the interdependence of the respective activities. The technical specification ISO/IEC TS 24748-6 shows this differentiation. It is also reproduced in this volume, but with more nuances.

The opposite of the verb **to integrate** is **to segregate**; segregate means to take apart, to dissociate.

Integration engineering

> Integration engineering is the set of activities that defines, analyses and executes integration across the system's life cycle, including interactions with other life cycle processes.
>
> (From ISO/IEC TS 24748-6 [ISO 24748-6]).

Integration is limited to prototypes or one-shot-systems.

Integration engineering encompasses all activities throughout the life cycle of a system-of-interest that are linked to the integration of the implemented components to form this system. These activities include:

- the definition, preparation and performance of the assembly of the implemented components of the system to form aggregates until the system as defined is obtained;
- the definition and performance of verification actions applied to aggregates to check the assembly, focusing in particular on interfaces;
- the definition and performance of verification and validation actions applied to the aggregates to check conformance of the assembly to system requirements, architecture and design;

- the integration of the formed system into its context of use (environmental context); integration is recursive and therefore applicable to the integration of the system-of-interest into the next level of the system structure.

Integration engineering also includes the activities of other technical processes that influence and/or constrain, guide, serve and enable the execution of the activities listed above. It also includes the integration of the system-of-interest with interoperating systems and enabling systems (a set of facilitating products and services) that support the integration of the system-of-interest.

The management of all these activities is part of the overall project management of the system-of-interest.

Notes:

The system-of-interest can be composed of layers of components. A component may be considered as a system or as a non-decomposable element. The system-of-interest is the highest abstraction level in the decomposition into systems levels.

While the system-of-interest is the highest level of abstraction in the decomposition of a particular system structure, this system itself can interact with other systems within one or more systems-of-systems. This perspective is necessary to facilitate the integration of the system-of-interest (in its environment) with interoperating systems and enabling systems.

Integration versus mass production

Integration of the system is part of the development effort related to the realization of prototypes (for validation purpose), or one-shot systems. The integration activity differs from that of the mounting of components / end products on a production or manufacturing line to obtain products of small, medium or large series.

For mass / series production, an assembly line does not necessarily use the same assembly order of the implemented components of the system as it is done for the prototypes within the integration process.

The integration of prototypes composes (creates) systems, using aggregates (see definition next paragraph), in order to verify and possibly to validate these aggregates and their interfaces, almost separately - see section 6.1.2 - Integration by layers of systems.

Mass / series production is of little interest to systems as defined in engineering. It focuses on the grouping of the implemented components in order to optimize the production time and effort. Nevertheless, the integration of a prototype provides relevant information to establish manufacturing and control files of the serial product, and to engineer a production line that replicates the prototype; in particular about the order of assembly of the implemented components of the system.

Notion of aggregate

> An aggregate is a composition of several "implemented system elements" (implemented components) of the system, which are assembled, on which a set of verification and/or validation actions is performed (From ISO/IEC TS 24748-6 [ISO 24748-6]).

The integration of a system is based on the notion of **aggregate**. An aggregate is a set of two or more implemented components of the system and their interfaces, as defined in the description of the architecture and design of the system. An aggregate has a functional consistency that allows the execution of verification actions and possibly validation actions. Each aggregate is characterized by a configuration of components of the system assembled physically, and by the configuration status specific to each component.

During integration, an aggregate does not necessarily represent a system as defined in the physical architecture or in the hierarchical decomposition of the system-of-interest. In order to efficiently integrate and validate the system, different aggregates may be constituted temporarily according to the integration techniques or methods used to define the integration, verification and validation strategies - see section 6.1.3. The validation strategy is particularly concerned, because the validation of the implementation of certain requirements is not possible using the complete system, for example, due to security constraints, physical or economic constraints. This item is addressed by forming temporary aggregates to verify conformance to the concerned requirements.

Notion of interface

> An interface is a set of logical and/or physical characteristics required to exist at a common boundary or connection between components of the system (From ISO/IEC TS 24748-6).

An **interface** is a concept. Any component of the system, which binds two components of the system, can be considered as an interface from an architectural perspective.

An interface generally includes two aspects:
- a **logical** aspect, i.e. the input-output flow and the function that carries this flow,
- a **physical** aspect, i.e. the technological physical link, which transports the input-output flow.

A **logical interface** consists of an input flow or an output flow or a bi-directional flow between two functions of the system, so that these functions may exchange material, energy and/or information.

A **physical interface** is a physical link, or port, that binds two components of the system-of-interest, or that connects a component of the system-of-interest with an element external to the system-of-interest. A physical interface may be considered a component of the system.

The definition of interfaces is an intrinsic part of the architecture and design of the system; it is essential to the success of integration.

Interfaces are common failure points in complex systems. These are the places where independent systems, or components of the system (not necessarily made of the same technology) meet and exchange with each other. This is why architectural and design definition activities and decisions must take into account how physical integration will be performed (assembly of system components and verification of the assembly).

6.1.2 Integration approaches

System integration *versus* system architecture and design

The term **system integration** covers activities related to the assembly of the implemented components of the system in order to obtain the corresponding final product or service or organization.

The integration of a system pre-supposes that architecture and design activities have been performed upstream, in particular to define the components and their interfaces; these may pre-exist, be re-used, or be specifically developed.

Based on stakeholder requirements and system requirements, the definition of the system architecture deals with the structure and composition of the components and their interfaces. The design of the components deals with the detailed description of the characteristics and the necessary technological enablers for their implementation and interfacing.

Scalability, maintainability, interoperability, availability, portability, etc., are stakeholder concerns that are addressed during the definition of architecture, to equip it with architectural characteristics such as modularity, standardized interfaces, encapsulation, etc.

Thus, integration activities consist, first of all, of defining the relevant order of assembly of the components of the system and defining the verification actions in relation to the characteristics of architecture and design. Next, comes the performance of the assembly of the components and the execution of the verification and/or validation actions.

Nevertheless, it is common to develop, improve or re-manufacture components without defining their order of integration into the final product. The order of assembly of the components must be flexible: different sub-assemblies (aggregates) must be able to be integrated into different sequences, though there may be a critical path for the supply and production of components. This consideration of the critical path for supply or production often makes it possible to identify integration constraints that influence requirements, architecture or design (especially with regard to interfaces). These constraints must be incorporated into system requirements, or in the definition of architecture or design.

As mentioned earlier in the section dedicated to interface, design and architecture related definitions and decisions must take into account how integration will be performed (assembly of components and verification of the assembly).

> Whatever the life cycle of the system, the integration of the components is based on the definition of architecture and design.

Integration scenarios

Sometimes, the term **integration** is used more restrictively, and consists of integrating pre-existing components. This may be the case, for example, with industrial practices that consist of selecting pre-existing products or COTS (Components Off-The-Shelf not necessarily initially defined to work with others) to be incorporated into a given system at a particular stage of its life cycle (e.g. in use).

Three examples of cases are given below, which show the importance of defining the architecture of the system, beyond the sole consideration of physical integration.

Planned integration

This is the classic approach used when developing a system, in which the components are defined before being physically integrated. These components have been defined or identified as COTS, when defining the architecture, to perform functions that are allocated to them in order to satisfy certain system requirements. These components may be developed specifically for the system-of-interest; they may be pre-existing and reused as is, or they may be modified. They may also be COTS developed and/or updated by external parties. The selection of components is an architectural and design decision. Planned integration may be achieved at design or build time; it may also be achieved through reconfiguration during operational use.

Evolutionary integration (evolution, extension, adaptation)

This is the approach used for an existing system, which is in service (in its utilization stage), and which must interact / interoperate with another one, existing or not. Integration cannot be performed without definition activities upstream; because this case can simply be regarded as a request, which includes specific requirements for interoperability, extension or adaptation. In particular, some architectural studies and evaluations have to be performed, such as interoperability analysis (protocols compatibility, interfacing capability), integrability analysis (i.e. feasibility of functional exchanges and physical connections), and verification and validation feasibility. These analyses are necessary inputs to the definition / modification of the architecture, in order to either, do nothing, modify the interfacing components, or add new components. Nevertheless, in order to perform these analyses, it is assumed that there are engineering data (documentation) of the concerned systems or components. Otherwise, a reverse engineering step must be performed to characterize the interfacing systems or components.

Capability extension integrating existing components or COTS

This case corresponds to a decision to develop a new service or functionality of an in-service system (in its utilization stage) by incorporating a pre-existing component or a COTS. This case is generally very constrained by a fast and low effort, a low-cost achievement and limited options. Moreover, these components can be significantly complex. This case can be considered as an extension request that includes specific requirements. Architectural studies and evaluations must be performed upstream in the light of these requirements. These studies include interfacing feasibility analysis, interoperability analysis, integrability analysis (feasibility of functional exchanges and physical connection), and feasibility analysis of verification and validation. An architecture based on a plug-in technology principle, or a strong modularity property, or on a service-oriented architecture (SOA), may be able to respond partially or totally to this situation. With such architectural characteristics, integration is facilitated, although it may not be visible immediately.

Notes:

The choice of COTS is generally motivated by strong time-to-market constraints and high level of technology readiness of COTS. The use of COTS constrains the architecture of a system and its integration (pre-existing interfaces). Deciding during the development of the system to use COTS, or a built in-house solution, facilitates the identification of appropriate interfaces and the definition of adequate verification and validation actions.

Never consider that a system-of-interest is **frozen** in its environment (or context of use). This environment may change over time; so, the system-of-interest must adapt to new operational conditions by adding or modifying capabilities (functionalities and performance). In this series of books, all types of systems are considered; including evolving and adaptive systems with dynamic self-configuration capability. This means that all methods, techniques, processes and approaches apply to any system, in particular that the integration process and all related processes can be used at any time in the life cycle of the system-of-interest.

Integration by layers of systems

The definition of a complex system (left part of Figure 69) is done progressively in successive layers of abstraction; each layer corresponds to a physical architecture composed of systems and non-decomposable components. The integration of the complete system (right part of Figure 69) consists of following the inverse path of the composition layer by layer. This general principle of integration is implemented and complemented as follows:

- In a given layer, the integration of the implemented components is carried out on the basis of the **physical architecture**, as defined during the system definition. But the **logical architecture** is also used as a basis for selecting components to be integrated that together perform a functionality of the system (functional chain). This leads to modifying the integration order of the components. The notion of **aggregate** has been introduced to allow flexibility in the integration strategy.

- One of the objectives of the system-of-interest integration is to gradually validate the complete system, by forming aggregates, so that certain critical components, or certain specific characteristics that require partial integration, are verified and validated as soon as possible. In this case, the integration method called "**criterion-driven integration**" may be used - see section 6.1.3.2 Integration methods or techniques.

Mixing these two ways (per layer and per functional chain, using appropriate integration methods) makes the integration and global validation of a system-of-interest efficient and optimized.

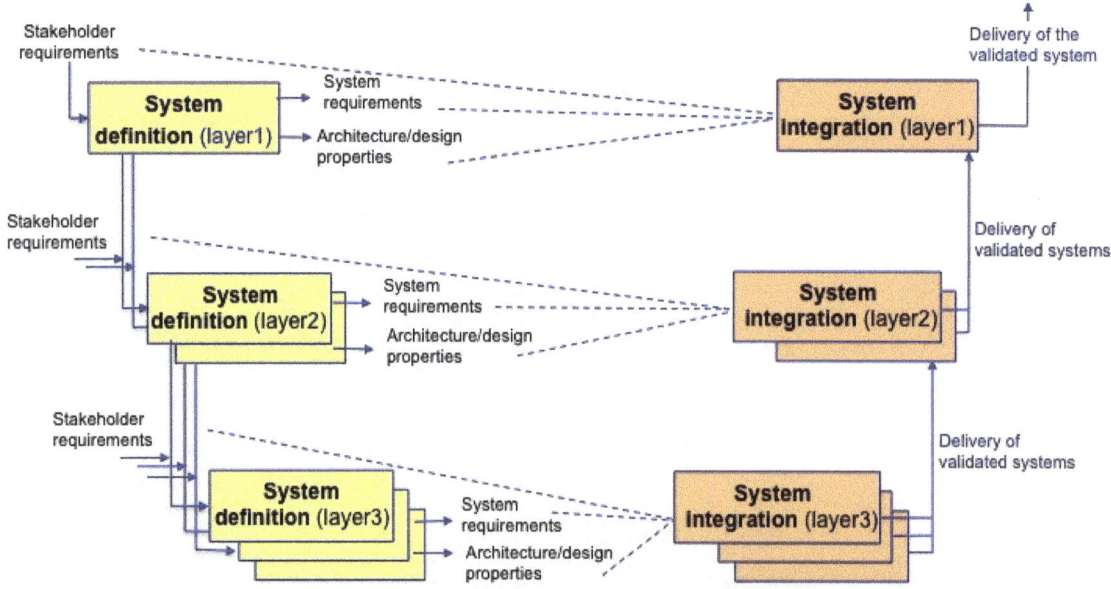

Figure 69 - Integration by layers of system

Many integrated systems are systems-of-systems; and integration begins with the combination of existing validated systems. In this case, the system-of-systems is the system-of-interest, and each existing system is a component within the system-of-interest.

Integration in the environmental context

Integration engineering also encompasses the integration of the system-of-interest into its environmental context, which may include physical, organizational, social and legal environmental conditions. When the system-of-interest has been fully integrated and functionally verified, it has to be incorporated into its operational environment and then transitioned into use.

The context of use (operational environment) can be considered as an upper level system for which the system-of-interest is a component. This incorporation follows the same approach, that is to say an integration step that consists of linking the implemented components of the system-of-interest with the elements of the context and verifying that the system-of-interest can operate under specified conditions.

The physical, organizational, social and legal environmental conditions, in which the system is intended to be used, are identified and described as applicable stakeholder requirements, which impact the corresponding applicable system requirements. These requirements are then taken into account as constraints during architecture and design definition to endow the system-of-interest with architectural characteristics and design properties. These properties and characteristics can have a significant impact on the effectiveness of the mission of the system-of-interest. These properties and characteristics may include, for example, thermal conditions, lighting, noise and spatial layout. Organizational and social aspects of the environment may include constraints such as work practices, organizational structure and attitudes. Safety aspects may include hazards or risks to life and property. Information security aspects can include expectations such as data integrity and vulnerability to certain threats.

Here are two examples of systems or products and their environmental conditions of use:

- Equipment used in a refrigerated warehouse is designed to take into account the need for operators to wear insulating protective gloves.

- A cash dispenser, which must be installed for use in an outdoor car park, is designed to take into account variable climatic environmental conditions (e.g. darkness, bright sunlight, etc.).

These considerations relate to **operational concepts** and **system requirements** that must be submitted to the validation activity. These engineering data can affect the integration strategy (on assembly and/or verification and validation).

6.1.3 Integration strategy

6.1.3.1 NOTION OF INTEGRATION STRATEGY

The purpose of the system-of-interest integration is to obtain a complete and integrated final system that conforms with the architectural characteristics and design properties as defined, and conforms with certain system requirements. Through the assembly of components, the objectives of integration are to verify that:

- all internal interfaces correspond with the characteristics defined in the architecture and design,
- all external interfaces meet the characteristics specified in the system requirements,
- all the functionalities are effective (executed) from end to end,
- all operational states or modes, dynamic aspects of operations are executed as intended,
- all degraded operations, overload conditions, dependability, safety, etc., are taken into account.

The response to this purpose and these objectives is provided by the integration strategy.

The integration strategy defines the activities that must be performed, where, when, how and by whom. Until obtaining the complete integrated system, that conforms with its architectural and design characteristics, integration activities include:

- the **reception** of each implemented component, previously verified and validated in isolation,
- the **physical integration** (assembly or connection through physical interfaces) of the implemented components to form aggregates,
- the **logical integration** of these aggregates (verification of the logical interfaces between the components through their functional, behavioural, temporal and other characteristics, as defined in the description of system architecture and design),
- in parallel, the **search** for architecture, design and assembly **defects**.

Note:
Anomalies, defects and non-conformances may be detected during integration; changes to the system are not included in the integration activities; a return to architecture and/or design or towards the technological realization is necessary.

Before performing any integration activity, the integration strategy has to be defined. The definition of the integration strategy is based on the architecture of the system and is based on the way this architecture has been defined. The strategy is defined in an integration plan that describes the configuration of the aggregates, the order of assembly of these aggregates (using appropriate integration techniques or methods - see section 6.1.3.2), in order to carry out effective verification and validation actions (e.g. inspections and/or tests). The integration strategy must be developed in coordination with the verification and validation strategies and the selected integration techniques or methods.

Based on the project objectives, such as delivery time, project cost, system safety and security, the integration strategy is defined, taking into account criteria such as integration deadlines and costs, risks, availability of resources and skills.

The integration strategy should be **as flexible as possible**; for example, in the case of large systems, in order to reduce delivery delays, the strategy is adapted and updated when stakeholders require the commissioning of a capability before full operation capability is available.

6.1.3.2 INTEGRATION METHODS OR TECHNIQUES

Here are some integration methods or techniques. The terms **driver / launcher** and **stub / cap** are used hereinafter

Driver / launcher: an external dynamic component or simulator that activates the execution of functions of an aggregate.

Stub / cap: an external static component or simulator that replaces one or more missing components in an aggregate and that provides predetermined results to enable the execution of functions of the aggregate.

Global integration

PRINCIPLE

Also known as big-bang integration. All delivered implemented components are assembled in one step; after having been verified and validated individually, the components are integrated concurrently.

ADVANTAGE

There is no need for driver / launcher, or stub / cap components to simulate absent components for verification purposes.

DISADVANTAGE

Interface errors are detected late. Detection, diagnosis and location of residual errors / defects are technically difficult, costly and inefficient.

APPLICATION

This method should be reserved for simple systems, with few interactions and some components; where technological risk is low.

Integration with the stream

PRINCIPLE

The delivered implemented components, after having been verified and validated individually, are assembled as they become available.

ADVANTAGE

This method allows starting the integration quickly.

DISADVANTAGE

Requires a large number of drivers / launchers, or stubs / caps to simulate absent components. A method that is not very efficient with respect to technical, means and cost aspects, because it requires a modification of the integration plan every time a component is integrated.

APPLICATION

Should be reserved for well-known and controlled systems, without technological risk. To minimize the risks, the method recommends planning the availability of the most critical components first.

Incremental integration

PRINCIPLE

In a predefined order, one or a few components are added to an already integrated aggregate (previous increment).

ADVANTAGE

Allows early detection of interface errors; makes it easy to locate the residual errors / faults of the new increment. The new errors detected are due to localized defects in newly integrated components or newly used interfaces.

DISADVANTAGE

Requires drivers / launchers, or stubs / caps to simulate missing components, and many test cases due to successive environments.

APPLICATION

The method can be used by any type of system architecture.

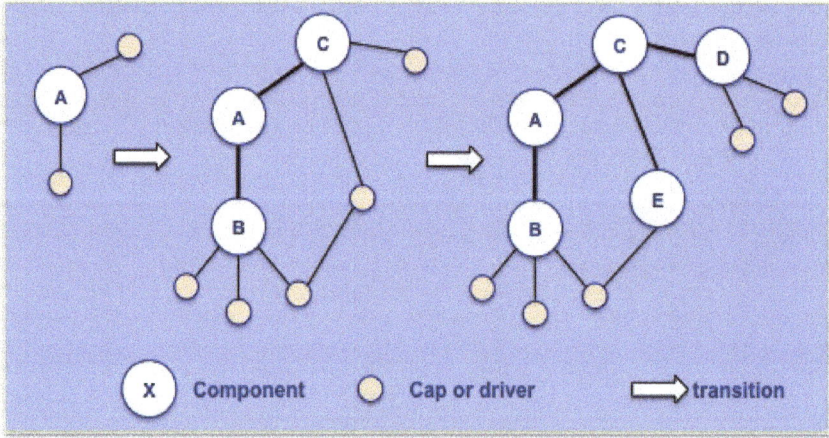

Figure 70 - Incremental integration

Subsets integration (aggregates)

PRINCIPLE

The components are assembled by subsets (a subset is an aggregate), and then the subsets are assembled together.

ADVANTAGE

Parallel integration of aggregates or (sub) systems is possible; the delivery of partial products or services is possible.

DISADVANTAGE

Requires a system architecture defined with (sub) systems; method better suited to (sub) systems with less complex interfaces.

APPLICATION

The method applies to systems architectures composed of (sub) systems; it applies well to products or services comprising relatively independent or low coupling functional or transactional chains.

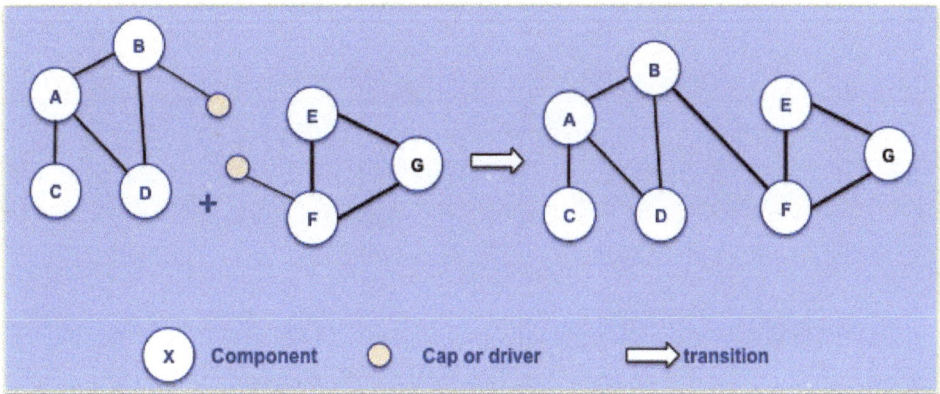

Figure 71 - Subsets integration (aggregates)

Top-down integration

PRINCIPLE

The integration starts with the components related to the overarching framework of use of the system-of-interest (generally components close to mission or users' vision and operations - when there are users and operators in the system-of-interest); it continues with components that perform functions / transformations of the core of the system; it ends with components that perform detailed functions or functions close to acquisition of signals or measurements (from sensors) and distribution of physical commands (onto actuators).

ADVANTAGE

Early availability of a system skeleton that allows easy detection and location of architectural errors / defects; definition of test cases close to reality; the method does not need many drivers / launchers.

DISADVANTAGE

Many stubs / caps to create; it is difficult to verify efficiently and thoroughly the components and interfaces of the lowest levels from test cases used for the high level (close to operators or users).

APPLICATION

The method is mainly used in software intensive systems, or in systems with human operators or users who pilot or interface with command-control (sub) systems.

Note:

A particular implementation of this method is the **onion ring approach**: integration begins with a kernel of the system or a backbone layer; then this kernel is used as a pilot / launcher to integrate other components. The kernel may include one or several functionalities shared by, or common to, a certain number of components.

Chapter 6.1. Integration: Concepts and approaches

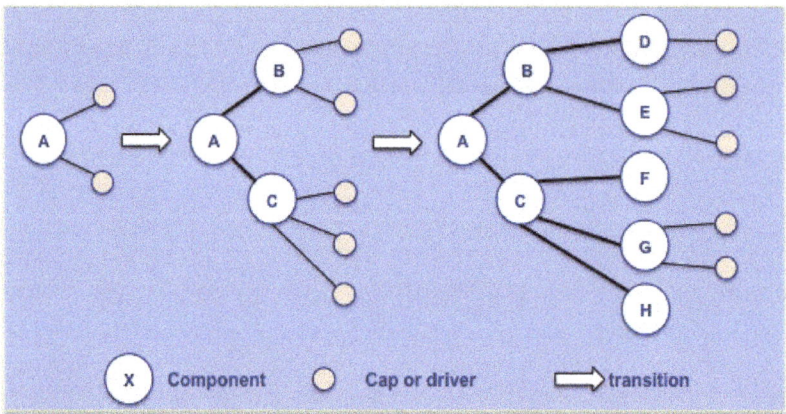

Figure 72 - Top-down integration

Bottom-up integration

PRINCIPLE

The integration starts with components that perform detailed functions or functions close to the acquisition of signals or measurements (from sensors) and distribution of physical commands (onto actuators); it continues with components that perform functions / transformations of the core of the system; it ends with components related to the overarching framework of the system-of-interest (generally components close to mission or users' vision and operations).

ADVANTAGE

No or few stubs / caps to create; early detection of errors / defects; test cases to check interfaces and functions are easy to define.

DISADVANTAGE

Lower-level components are solicited a lot by higher-level test cases; absent components from the upper levels are replaced by drivers / launchers to be created.

APPLICATION

It is almost the only way to integrate hardware components; it also applies to software intensive systems.

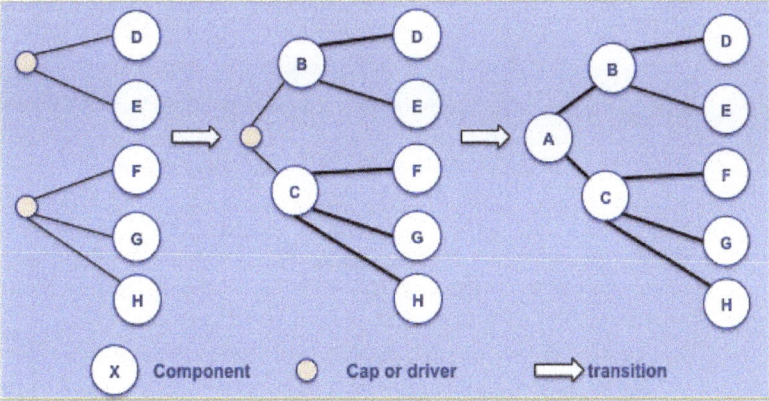

Figure 73 - Bottom-up integration

Criterion driven integration

PRINCIPLE

The most critical implemented components with respect to the selected criterion are integrated first. The criteria are derived from system requirements and are generally related to technical risks; these criteria relate to dependability, security, complexity, performance or effectiveness, usability, technological innovation, etc. Aggregates are defined around components that present technical risks.

ADVANTAGE

Allows for early and intensive testing of critical components; early verification of design and architectural choices.

DISADVANTAGE

Drivers / launchers and stubs / caps are often difficult to define and implement.

APPLICATION

The method applies to any system architecture that includes critical components.

Mix integration method

The selection of integration techniques or methods depends on several factors, such as the type of components, their criticality, delivery time, delivery order, technical risks, constraints, verification strategy, validation strategy, project type and organization. Each technique or method of integration has advantages and disadvantages that must be considered when defining the integration strategy.

Usually, a mixed approach of the different techniques or methods listed above is selected, in order to optimize the integration work and to adapt the process to the system under development.

The integration activity is not a phase of the project fixed in time; it can also be performed dynamically during the operation of the system through self-configuration arrangements of the system-of-interest architecture. The top-down integration method can be used in this case; the presence of stubs / caps enables to progressively build a system driven by requirements, such as availability, scalability and adaptability.

6.1.3.3 ASSEMBLY TECHNIQUES

The purpose of **assembly** is to obtain the various subsets or aggregates defined in the integration plan. It can be done in **sequence** (one component after another), or in **parallel** (several groups of components assembled in parallel) - see illustration in Figure 74.

The assembly of complex subsets must be detailed in the integration plan.

During assembly, it is possible to perform some preparation and/or finishing tasks that require the simultaneous presence of the concerned components. For example, painting two pieces together after assembly, drilling and tapping two facing pieces, tuning software settings once loaded on the vehicle. These operations must be planned and cannot be performed on the isolated components. Under no circumstances should they involve architecture and design changes.

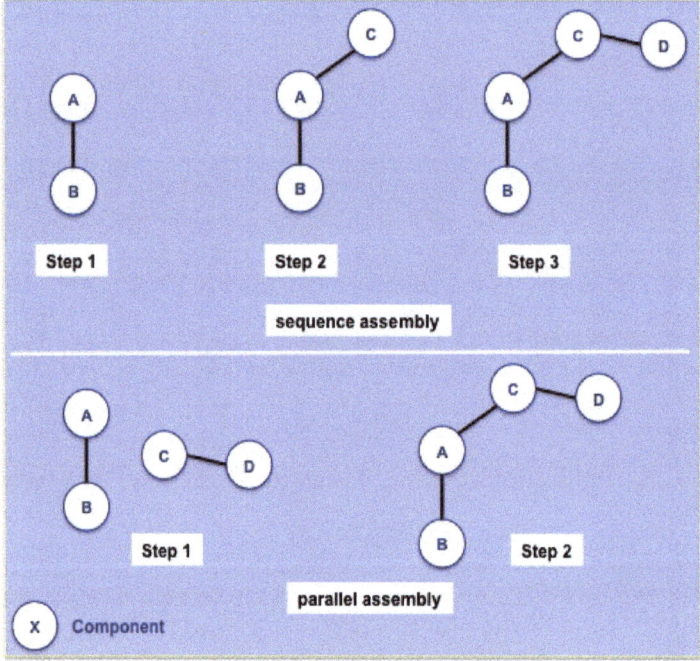

Figure 74 - Assembly techniques

6.1.3.4 INTERFACES ANALYSIS

In order to make the integration strategy more efficient, the order of integration of the components can be done step by step through the analysis of the interfaces. The coupling matrix (or N^2 representation) is a typical interface analysis method.

The N^2 diagrams are used in the system-of-interest definition to identify and define the logical interfaces between the functions. They usually involve functions, but they can be used with components. These are tools commonly used by integrators to define the order of integration of components and verification of interfaces. N^2 diagrams and coupling matrices are equivalent representations.

Coupling matrices allow defining aggregates of components as well as the verification of the interactions between the components. During verification, they also make it easier to locate potential remaining errors in the system.

A coupling matrix presents the components on the diagonal of a square; the other cells of the table identify the presence or absence of an interface between the components; the upper right corner contains the interface direction from Ex to Ey, while the lower left corner contains the interface direction from Ey to Ex. An interface cell may simply indicate that an interface exists, or may contain the number of interactions or interfaces that connect components. Figure 75 shows only the presence (or not) of an interface. Inputs are identified by an x in the corresponding column, either above or below the component. For example, the inputs of E3 come from E1, E2 and E6. The outputs are identified by an x in the corresponding line, either to the left or to the right of the component. For example, the outputs of E3 go to E1 and E6. A blank cell indicates the absence of an interface between the components.

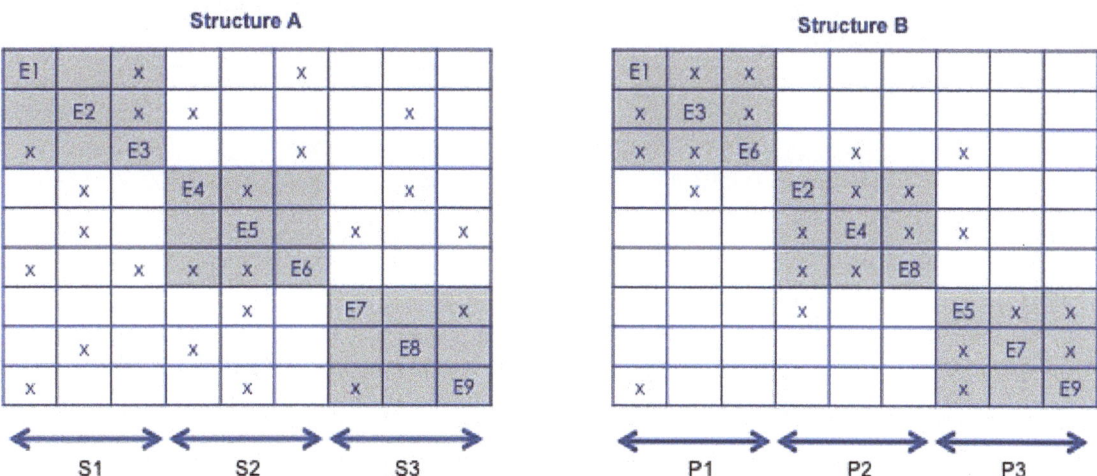

Figure 75 - Example of coupling matrices to define integration aggregates

Figure 75 presents two possibilities for the definition of aggregates. Structure A, on the left side of the figure, includes three aggregates S1, S2 and S3; these aggregates have many interfaces and interactions between them that require defining at least 16 verification actions. In case of anomalies in the global functioning, it is difficult to locate the errors.

Structure B (obtained by moving the columns of structure A) on the right side of the figure includes three aggregates P1, P2, P3; each aggregate has only 2 interactions that lead to 2 verification actions; the total number of verification is 6: 2 between P1 and P2, 2 between P2 and P3, 2 between P1 and P3. In the event of an anomaly, errors are easy to spot. The structure B increases the cohesiveness within the aggregates and reduces the coupling between them.

Coupling matrices are also an aid to the definition of the test cases to check the outputs from the inputs of the aggregate, and to exert the flows through the components between the inputs and the outputs. Figure 76 shows a representation of the coupling matrices that includes the inputs and outputs of the aggregate.

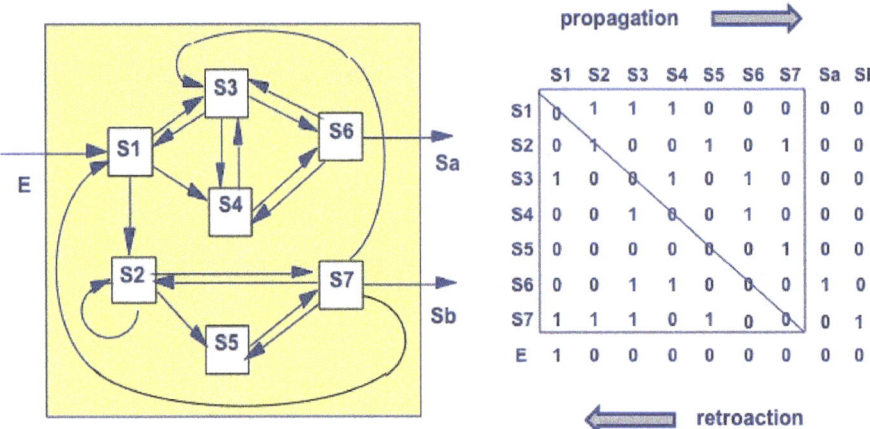

Figure 76 - Coupling matrix to define test cases for input / output flows

6.1.3.5 VIRTUAL INTEGRATION, SIMULATION

As soon as the design of the components and the system architecture are defined, virtual environments, modelling activities, mock-ups, or other practical tools can be used to model or simulate the assembly of the designed components. These activities provide an early check on the feasibility of the assembly, the definition of necessary checkpoints, and the possibility of simulating the connections of the interfaces.

Performed before physical integration, these activities allow the discovery of late integration errors in development, and to mitigate the risks more easily. The result of these activities may lead to the incorporation of arrangements in the architecture and/or to modify the design of the components and/or their interfaces.

6.1.4 Relationship of integration with verification and validation

Integration and verification

Integration activity is not limited to assembling implemented components. When one component is assembled with another, the interfaces are checked more or less simultaneously. Several verification actions are often required before assembling or connecting components; otherwise accidents may happen. Examples of verifications:

- inspection of the dimensions of the physical faces of each component to be connected,
- measurement of the electrical voltage required between the delivered and used power,
- examination of the position and/or polarity of pins,
- verification of the compatibility of interfacing protocols in the software domain,
- inspection of software code to check that interfaces are implemented identically in interfacing components.

The verification process is widely used by the system integration activities. As a reminder, the purpose of verification is to detect the maximum number of errors or defects that the system may include, and this before delivery. Due to the number of items to be verified, time and cost could exceed the expected budgets; this is why a verification strategy has to be established at the same time as the integration strategy.

Integration and validation

As explained previously, validation of conformance with certain stakeholder requirements and/or system requirements is not possible using the complete system due, for example, to safety, security, physical or economic constraints. Temporary aggregates are thus formed in order to establish conformance with the concerned requirements.

The validation process is also widely used by system integration activities. As a reminder, the purpose of validation is to obtain the maximum confidence in the use of the system, before its final delivery (i.e. transfer of property from supplier to acquirer). Due to the number of requirements, in which conformance has to be checked, time and cost could exceed expected budgets; it is for this reason that a validation strategy has to be established at the same time as the integration strategy.

Testing levels

Component level testing

Throughout the integration of the system, verification and validation actions of *test type* are performed. Integration of the system assumes that **component level tests** have been successfully completed before the components are delivered, in isolation for each component. As a reminder, component level tests include *functional tests* (nominal and non-nominal behaviour; interfaces), grey box *structural tests* and glass box *structural tests* (own resources, intermediate states) - see section 5.3.4.1. The comparison reference is essentially the design description document.

Integration level testing

Integration level tests are performed on aggregates (assembly of previously verified components one by one) and then on aggregate sets until they form the complete system-of-interest (product or service or organization). Several steps are therefore necessary to fully integrate the system-of-interest.

The comparison reference is essentially the description of the system architecture. The objective is to verify the assembly of the components according to the architecture, and the interfacing mechanisms.

Functional tests are used to verify the functions performed by the subset of the architecture, as defined by the aggregate, and the interfaces with the other aggregates, external products and external systems.

The grey box *structural tests* verify that:

- components interact correctly when activated together,
- flows are exchanged correctly between the components of the aggregate,
- dynamic interfaces mechanisms and supervisory control devices (if any) work correctly: task synchronization, shared resources, communication protocols, interrupt management, etc.

Chapter 6.1. Integration: Concepts and approaches

Validation level testing on industrial site

These tests involve verifying the conformance of the complete implemented system (product, service, organization), essentially to system requirements. These tests are *functional tests*; it is too late to perform *structural tests* (this is in the scope of integration level testing).

The functional tests, for this level, are defined from:

- functional requirements,
- interface requirements (with outside of the system),
- operational requirements related to utilization (transfer for use, maintenance, logistic support, etc.),
- effectiveness requirements (resources for time and volume).

Practically, the tests can be performed for each functional chain separately (functionality + effectiveness + ergonomics + safety), then functional chain sequences.

Examples of validation tests on industrial site:

- for each function or transaction:
    - nominal modes
    - degraded modes
- for interfaces:
    - exchanges with external objects
    - communications (messages, coupling with other products)
    - information flows
    - man-system interfaces (exit status, screens, error messages, etc.)
- operational dialogue and man-machine interface:
    - screens kinematic, function keys
    - path of dialogue graph
    - inputs access, dialogue interruption
- effectiveness: precision, loading, response time, transmission time
- dependability: integrity, safety

Validation level testing on operational site

These tests include the verification of the purpose of the system-of-interest; i.e. the capability of the system-of-interest (product, service, organization) to participate in the achievement of the mission of the upper-system (system in which this system-of-interest is immersed). The comparison reference consists essentially of stakeholder requirements (or system requirements of the upper system).

The test cases of this level mainly concern:

- operability (in near-real-world operation, in nominal mode limits, in degraded mode due to component failure),

- robustness (availability of peak load resources, vulnerability to threats),
- dependability (global availability, safety, survivability or resilience),
- installation and transfer for use.

Examples of validation tests on operational site:

- loading tests: large volume of inputs flows or stimuli in a short period of time (at critical moments in available resources, the system uses mechanisms rarely solicited in nominal modes)
- utilization tests related to human factors:
  - Interfaces consistent with the level of qualification of operators (dialog and error messages)
  - ergonomics: conventions, format inputs, confirmation of actions, excessive query options, acknowledgment of receipts
- safety and security tests:
  - protection mechanisms (immunity)
  - intrusion mechanisms
- configuration tests: types and number of communication lines, unavailable services, migration
- installation tests: configuration of devices, resources, services delivered, installation conditions and installation handbook, settings and calibrations
- operation, maintenance tests: input flows, incident recovery, implementation handbooks

Integration strategy effectiveness

The effectiveness of the integration strategy is to minimize workload, time and risk, in order to achieve a complete integrated system-of-interest capable of operating within its context of use. For this, it is recommended to:

- define the appropriate order for assembly of components using defined procedures,
- minimize the number of verification actions to verify the physical interfaces before and during the assembly or connection of the components in an aggregate,
- minimize the number of verification actions to verify logical interfaces after assembly or connection of components, as well as the functionalities of the corresponding aggregate,
- minimize the number of validation actions to verify certain functionalities and system requirements in the corresponding aggregates.

The effectiveness of the integration of the system-of-interest is obtained by considering, at the same time, the definition of integration, verification and validation strategies. It consists of defining and executing a minimum of necessary verification and validation actions, while ensuring the maximum confidence that the system will operate as planned once the integration has been completed. The definition of the strategy includes potential risk studies if certain verification and validation actions are withdrawn.

6.2 Process approach

This chapter describes:

- ♦ The location of the process in the development cycle - section 6.2.1
- ♦ The purpose, inputs and outputs of the process - section 6.2.2
- ♦ The activities of the process - section 6.2.3
- ♦ The ontology elements used - section 6.2.4
- ♦ The main artefacts produced by the process - section 6.2.5

6.2.1 Location of the process

Figure 77 shows the position of the integration process among the technical development processes of the system-of-interest. From the viewpoint of the partition of activities into the life cycle processes, it is to be noted that the integration process intensively calls upon the verification process and the validation process. Depending on the projects and organizations, it is possible to include or exclude the final validation activity (on industrial site and/or operational site) from the integration process.

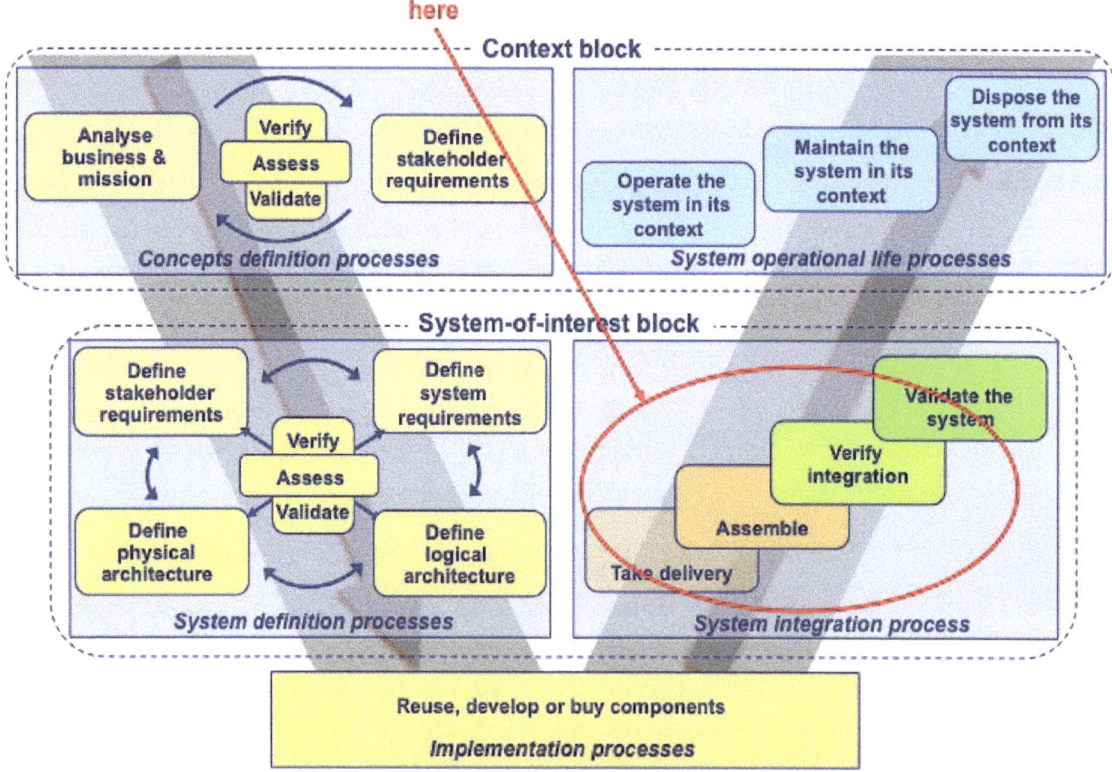

Figure 77 - Location of the integration process in the development

6.2.2 Purpose, inputs and outputs of the process

Synthetic and generic expression

The purpose of the integration process is to assemble the implemented components, in order to obtain the product, service, organization that conforms with the elements of the system architecture, and the associated design properties.

The aim of integration is to obtain the product, service, or organization ready for final validation and then for transfer for use or production.

The integration process applies to each level of decomposition of the system-of-interest, except at the lowest level of implementation of the technological components. It is used iteratively starting from a first aggregate of components, up to the complete system. The result of the last iteration of the process forms the system-of-interest.

Note: For integration activities, the term *system* should not be used, as it is a matter of realizing the product, service or organization that supports the system. Nevertheless, the denomination is common and is often used in this book.

Inputs

The inputs of the process are:

- the implemented components, and previous integrated systems (iteration of the process),
- the description of the system architecture,
- the components and previous integrated systems design properties,
- the system requirements,
- the Assembly tools and/or means.

Outputs

The outputs of the process are:

- the integration plan, which includes the system integration strategy,
- the successive Aggregates (up to the integrated product, service, organization),
- the Assembly procedures,
- the integration reports,
- the anomaly and/or non-conformance sheets.

Chapter 6.2. Integration: Process approach

6.2.3 Activities of the process

The activities of the integration process follow:

      A. Prepare integration

      B. Obtain assembly tools and means

      C. Take delivery of components

      D. Assemble components into aggregates

      E. Verify and validate aggregates

      F. Manage integration results

      G. Control the integration process

Figure 78 presents the activities of the process, the exchanged engineering data and the main artefacts generated by the process.

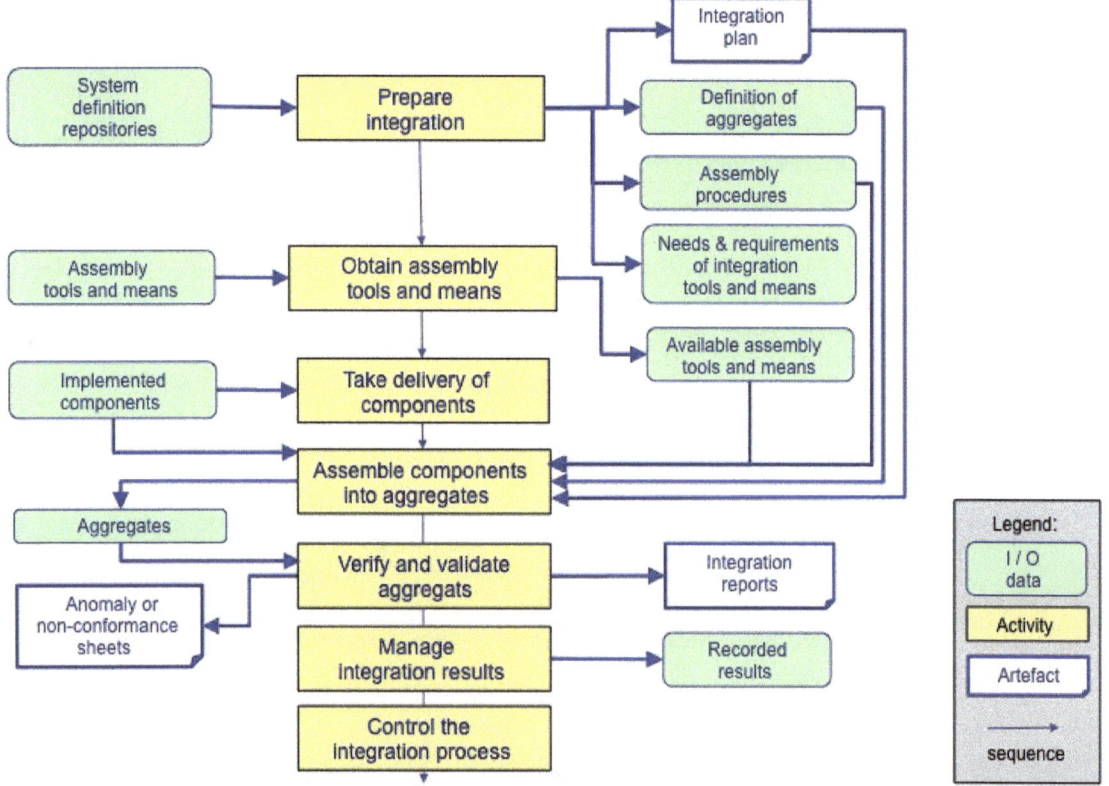

Figure 78 - Activities of the integration process

Details of activities

A. Prepare integration. This activity includes the following tasks:

1. Define the integration strategy. The integration strategy describes how integration has to be prepared and performed - see section 6.1.3.1. It is defined as follows:
    a. analyse the description of the architecture of the system, in particular the physical and logical models, in order to identify the **Components** and their interfaces:
         i.   first the components that perform the functional core of the system,
         ii. then the components that perform functions related or necessary to the functional core of the system,
         iii. finally, the components that interface with elements external to the system;
    b. analyse **System requirements** and architecture descriptions to define the order (priorities) of assembly of the components;
    c. study candidate integration techniques or methods that may be relevant to the type of system to be integrated - see section 6.1.3.2;
    d. study **Aggregate** alternatives and sets of aggregates that could predefine the order of assembly;
    e. analyse the verification and validation strategy (see section 5.1.5) to determine which **Components** and **Aggregates** are necessary to perform the selected **Verification and/or Validation actions**; this strategy contains the list of verification and validation actions that will be executed; this list is the result of trade-offs between what **must** be verified and/or validated and what **can** be verified and/or validated taking into account the multiple constraints, such as cost, deadlines, feasibility, safety, security, etc., and risks if certain verification and validation actions are withdrawn;
    f. perform hazard, safety, security and risk analyses to avoid unintended or undesired consequences that may arise during the execution of the integration or be generated by it; these consequences include the integrity of the system itself, the safety of operators inside or outside the system, integrators and environmental conditions;
    g. assess and select an integration technique or method (see section 6.1.3.2), or define a mix of them taking into account the previous studies;
    h. define the assembly order of the **Components** using the sets of aggregates previously defined, and select an assembly technique (parallel and/or sequential - see section 6.1.3.3);
    i. consider project parameters and constraints to minimize integration time, cost and risk; synchronize integration tasks with the project master schedule;
    j. document the integration strategy in the integration plan that indicates **where** (sites), **when** (schedules), **how** (procedures) and **who** (competencies) performs integration activities and tasks.
2. Define assembly tools and means. This task consists of identifying and planning the enabling systems, services and tools needed to perform the assembly operations. These facilities include infrastructures, assembly equipment, harnesses, training systems, simulators, measurement devices, safety elements, etc. In more detail:
    a. identify **Assembly tools** and means necessary for the construction of the aggregates according to the selected integration methods or techniques, taking into account the verification and validation techniques used in the verification and validation strategy;

      b. define the requirements of these means and tools, in particular the interface requirements with the **Components** to be integrated;

      c. plan the availability of the **Assembly tools** and means.

3. <u>Develop assembly procedures.</u> For each **Aggregate** as defined above, the **Assembly procedure** describes the sequence of assembly operations of the **Components** between them, the **Assembly tools** and the necessary skills, the **Verification and/or Validation actions** (inspections, tests) to run **before**, **during** and **after** assembly.

4. <u>Identify integration constraints or requirements of the system.</u> These are the constraints coming from the definition of the integration strategy. Some of these constraints must be incorporated into the **System requirements**, Architecture or Design. For example, accessibility, security for integrators and operators, interconnections needed for **Aggregates** and **Assembly tools** or means, and interface constraints. Constraints to be incorporated into the definition repositories may affect the use or operation of the system (such as check points to be incorporated into the system) and architectural and design characteristics (for example, the location of components, input-output flows or commands exchanged with operators or users, interfaces with harnesses or drivers / launchers and stubs / caps).

B. Obtain assembly tools and means. These are means and **Assembly tools** and **Verification and/or Validation tools** as defined in the integration plan and in the verification and validation plan. Depending on the means and tools involved, they can be bought, rented, developed, reused or outsourced. These means must be acquired, **verified**, **validated** and **available** according to the planned schedules.

C. Take delivery of components. This activity includes the following tasks:

<u>Unpack each component following the unpacking instructions</u>. **Components** are received from the suppliers, the purchaser, or removed from the storage provided for this purpose.

1. <u>Perform receipt checks.</u> For this you have to check:

      a. that the transport did not lead to degradation of the component,

      b. that interfaces conform with the requirements,

      c. that mandatory documentation is present (installation and user manual),

      d. that contractual documentation is present,

      e. the presence of the conformance document or proof of the verifications and validations,

      f. the presence of the configuration status document (identification, version, list of changes introduced in relation to requirements, lists of accepted deviations, etc.),

      g. the physical marking versus configuration status.

2. <u>Re-assemble each component with its accessories.</u> Follow the assembly procedure provided. The component must be handled in accordance with the health, safety, security and confidentiality rules provided.

3. <u>Accept or reject each component</u>. Components that do not pass the checks are identified as such and treated in accordance with the disposal procedures, or returned to the supplier. If accepted, they are identified as such.

D. Assemble components into aggregates. This activity includes the following tasks:

1. For each integration aggregate, gather the necessary elements for integration. These elements are described in the integration plan and in the assembly procedures, namely:

    a. **Implemented components** used in the composition of the **Aggregate**,

    b. **Assembly tools** and means, **Assembly procedures**,

    c. **Verification** and/or **Validation tools** and means, **Verification procedures**,

    d. the skilled integration operators.

2. Assemble the implemented components to form each aggregate. For this, the operators perform the following operations:

    a. connect the **Implemented components** to each other to constitute the **Aggregate** in the order prescribed by the integration plan and according to the **Assembly procedures** using the tools provided,

    b. add or connect the **Verification** and/or **Validation tools** to the **Aggregate** as defined in the procedures,

    c. perform any welding, gluing, drilling, tapping, adjustment, parameterization, painting, calibration, etc., operations; these finishing tasks simultaneously concern several components,

    d. check all assembly points as defined in **Assembly procedures** **before**, **during** and **after** their execution.

3. Document the integration report. Record events and assembly results.

E. Verify and validate Aggregates. This activity includes the following tasks - see also section 5.2.1.3 and section 5.2.2.3:

1. Check the correct connection of verification and validation tools.

    a. These tools are linked to the **Implemented components** of the **Aggregate** through check points; equipment such as probes, sensors, spies, plotters, specific analysers, etc., can observe the behaviour of the aggregate during the execution of the **Verification** and/or **Validation procedures**.

    b. The tools also include drivers / launchers or stubs / caps to simulate missing components or components outside the system.

2. Perform verification and validation procedures. **Before**, **during**, and **after** the execution of each procedure, check that the aggregate has the expected characteristics as described in the **Verification and Validation actions**: behaviours, functions, I/O flows, effectiveness, design properties, system requirements, etc.

3. Collect measurements and results. Throughout the execution of the **Verification and Validation procedures**, collect data, measurements, and results.

F. Manage integration results. This activity includes the following tasks:

1. Record assembly results. These results are contained in the integration reports. Add any anomaly or non-conformance sheets.

2. Record results of verifications and validations. This task is described in detail in sections 5.2.1.3 and 5.2.2.3, activity number C.

G. Control the integration process. This activity includes the following tasks:

1. Coordinate activities of the integration process with those responsible for the activities of the system definition processes (user or customer representatives, specifiers, architects, developers, designers) for the technical aspects of integration.

2. Coordinate the activities of the integration process with the System Development Project Manager for the timing and cost aspects of the execution of the Assembly procedures, for the procurement of means and Assembly tools or enabling systems necessary for integration, for the availability of qualified personnel, for the supply of resources, etc.

3. Coordinate the activities of the integration process with the Configuration Manager for Verification & Validation Configurations, anomaly or non-conformance sheets, repository, files, and object versions.

4. Update the Integration Plan (integration strategy, detailed schedules, etc.) as the project progresses.

5. Update the Assembly procedures and all other elements due to anomalies discovered during the integration execution.

6.2.4 Ontology elements

Elements

The integration process uses the main engineering meta-data indicated in the **Table 11**.

ELEMENT	DEFINITION AND ATTRIBUTES (examples)
Assembly procedure	An assembly procedure groups a set of operations executed together simultaneously or sequentially in a defined sequence / scenario.
	Identifier; Description; Duration; Unit of time; Comment
Assembly tool	Physical equipment and/or computer equipment used to perform one or more assembly procedures (test bench, simulator, harness, cap, launcher, measurement device, enabling system, etc.).
	Identifier; Description; Comment
Implemented component (also Integrated system)	Component (System element) that has been implemented (manufactured, assembled); it is identified by a serial number. Examples: software application, mechanical / electrical / electronic hardware component, ... operator role, procedure, protocol, etc.
	Identifier; Description; Comment
Integration aggregate	Composition of several "implemented components (system elements)" of the system which are assembled, on which a set of verification and/or validation actions is performed.
	Identifier; Description; Comment

Chapter 6.2. Integration: Process approach

ELEMENT	DEFINITION AND ATTRIBUTES (examples)
Verification or Validation configuration	A verification or validation configuration gathers physical elements (objects to be verified or validated, verification or validation tools or means) necessary for the execution of a Verification or Validation procedure.
	Identifier; Description; Comment
Verification or Validation tool	Physical equipment and/or computer equipment used to perform one or more verification or validation procedures (test bench, simulator, harness, cap, launcher, measurement device, enabling system, etc.).
	Identifier; Description; Comment

Table 11 - Main engineering meta-data related to integration

Relationships

The main relationships between the engineering meta-data are presented in Figure 79.

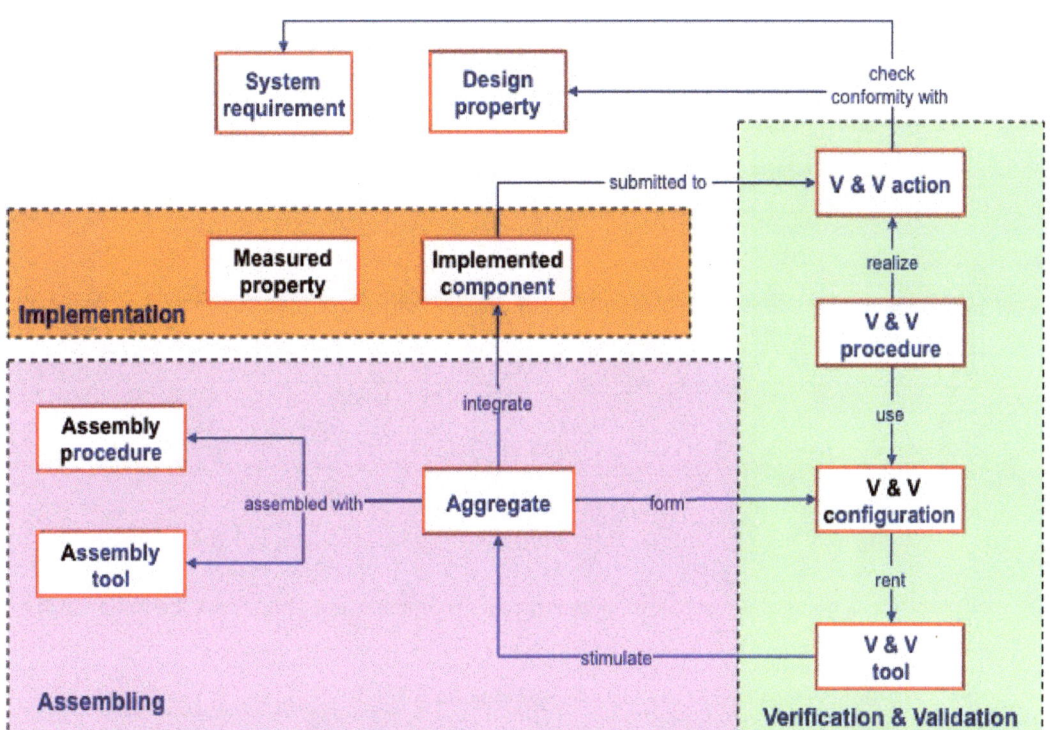

Figure 79 - Relationships between engineering meta-data related to integration

Utilisation of ontology elements

The meta-data defined above in Table 11 and their relationships in Figure 79 represent the skeleton of the activities and tasks of the Integration process. During the execution of tasks of the process the experts in systems engineering have to:

- Create as many as necessary instances of these meta-data: Aggregate, Assembly procedure, Assembly tool.
- Fill in attributes of these instances with values; the generic attributes represent characteristics of a meta-data; particular values for one instance particularise this instance among others.
- Instantiate the generic relationships between meta-data to establish traceability links between instances.

The main relationships to establish during the execution of this process are the following:

- an Aggregate integrates an Implemented component
- an implemented component is integrated into an Aggregate
- an Aggregate is assembled with an Assembly procedure
- an Assembly procedure assembles an Aggregate
- an Aggregate is assembled with an Assembly tool
- an Assembly tool assembles an Aggregate
- a Verification or Validation configuration is formed by an Aggregate
- an Aggregate forms a Verification or Validation configuration
- a Verification or Validation tool stimulates an Aggregate
- an Aggregate is stimulated by a Verification or Validation tool

Implemented by software applications such as Data Bases or Repositories or Spreadsheet tools, these relationships enable the generation of Traceability Matrices; these latter are useful for various purposes such as verification, impact analysis (in case of evolutions), justification, technical risks management, and consistency checking.

6.2.5 Artefacts - documentation

This process generates the following documents. Templates and Guidelines are provided in Annex.

- Integration Plan - see Annex chapter 8.9
- Aggregate Definition Sheet - see Annex chapter 8.10
- Assembly Procedure - see Annex chapter 8.11
- Integration Report
- Anomaly or Non-conformance Sheet - see Annex chapter 8.6

6.3 Practice

This chapter provides explanations, practices and recommendations related to integration:

- Components heterogeneity and interfacing complexity - section 6.3.1
- Notion of Integration Enabling System - section 6.3.2
- Interfacing with other development processes - section 6.3.3
- Defects searching and clearing - section 6.3.4
- Recommendations and FAQ - section 6.3.5

6.3.1 Components heterogeneity and interfacing complexity

Interfacing of heterogeneous components

One of the major difficulties in interfacing is the presence of components based on different / heterogeneous technologies. The following three cases illustrate this heterogeneity, and provide some recommendations:

- In the case of systems mixing IT (computers, peripherals) and control of industrial processes (sensors, actuators, converters, etc.), the verification of an interface should be done in several steps: first, the functionalities of the interface (flows exchanges between heterogeneous components), then its effectiveness (response time, volume of information), and finally its service qualities (availability, accuracy, integrity, etc.). For example, we verify that the interface provides correct results (functionality), then that the results are delivered within the expected time (effectiveness), and finally with the expected accuracy (measurement transmitted).

- In systems using technologies such as mechanics, electricity, chemistry, etc., verifications should be carried out on assembly aspects in order to investigate physical phenomena at the contact surfaces of the components; for example, tolerances of assembly, slack, friction, expansion, capillarity, etc. Each interface between components of different technologies may present a particular difficulty. Before starting the integration, it is recommended to study the interfacing of different technologies using models reduced to these aspects.

- When integrating higher-level abstraction (sub) systems, undesired **emergent properties** may appear; e.g. electromagnetic induction, interference, vibration, resonance, turbulence, instabilities, etc. These properties are generally revealed on subsets or aggregates of significant size. Diagnosis, localization and treatment of these properties are not simple:
  - the diagnosis and localization require the implementation of systematic means of detection;
  - the treatment usually requires a partial rework of the architecture and/or design.

As a consequence, the integration of **innovative** complex systems can be difficult to predict. That is why it is recommended to:

- set up a highly visible progress monitoring of integration activities,
- anticipate difficulties, especially potential emerging properties,
- establish an integration plan and a verification - validation plan flexible (allowing changes in schedule, order of assembly, order of verifications) and which contain time margins,
- to integrate sets by technologies of the same nature, if possible.

This last point is illustrated by the example in Figure 80. This figure should be read from bottom to top; it is a mixed integration method that combines the bottom-up integration method and the criterion-driven integration method (the criterion is technology).

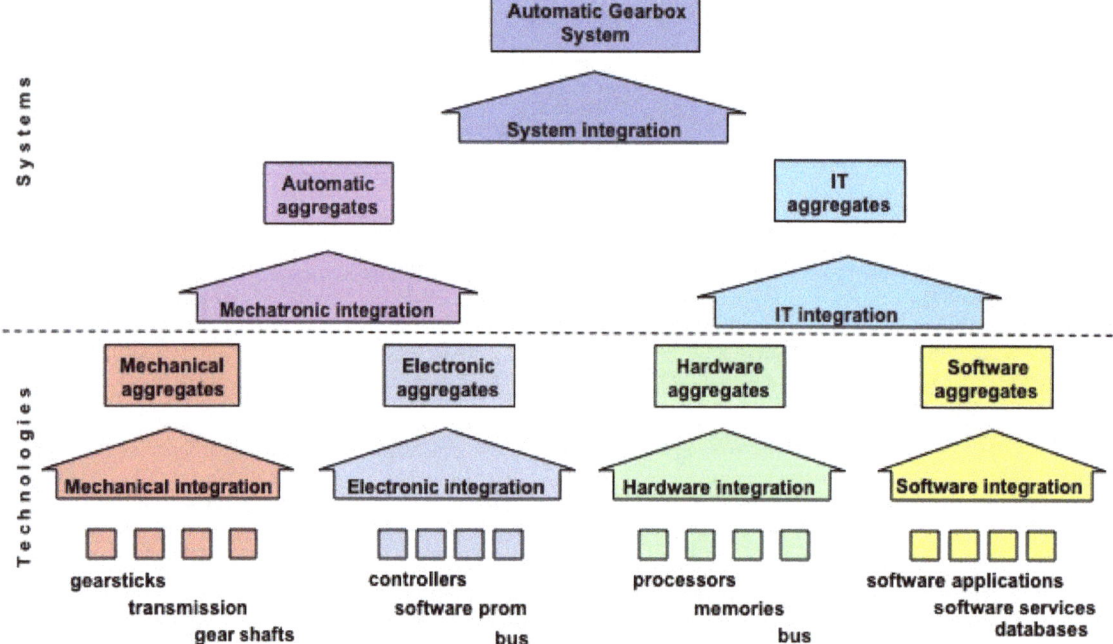

Figure 80 - **Integration by similar technologies** (read from bottom to top)

This example illustrates how technologies integration and system integration can be articulated within a development project.

This is to be compared with the transition from system engineering to technology engineering - see Chapter 6.5 of Volume 1 of this series, System Notion and Engineering of Systems [Faisandier1 2015]. As a reminder, in the overall system, each technological set must present a coherent whole, i.e. coherent architectural characteristics and design properties, and in relation to those of the overall system. During engineering / definition, technological architecture activities are carried out to build the cohesion of each set, using specific architecture and/or design definition processes - see Figure 81 (an end component may consist of one or more technologies).

During the integration of each technological set, the verifications and validations have for goal to check the respect and the coherence of the architectural characteristics and the design properties. These observations demonstrate, in particular, the respect of the requirements from which these characteristics and properties are derived.

Note:

This technology integration approach constitutes technological aggregates to verify the design of technological architectures (business views). Once these checks have been acquired, it may be appropriate to dismantle these aggregates to create others on the basis of integration by levels of systems - see Figure 69, section 6.1.2.

Chapter 6.3. Integration: Practice

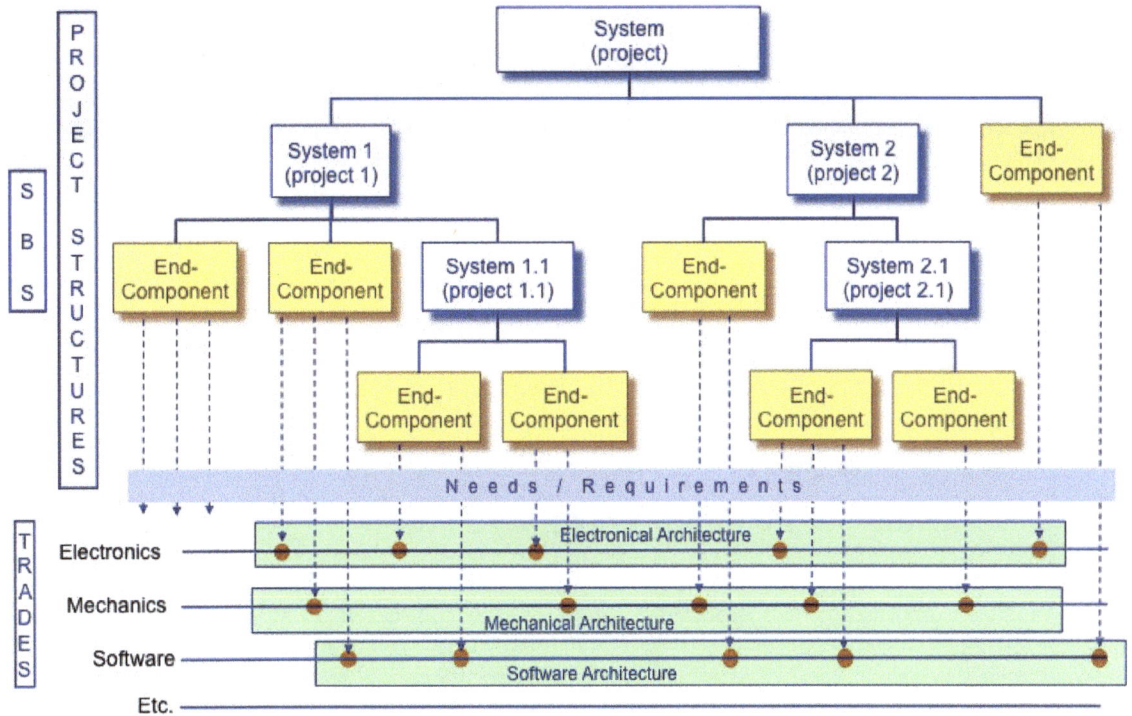

Figure 81 - Relationship between engineering of the system and engineering of technologies

Complexity of interfacing

With regard to interfacing in complex systems, a distinction can be made between static complexity (physical connection interfaces) and dynamic complexity (interaction of exchanges via interfaces).

Static complexity can be controlled by improving the **quality of coupling** (see Volume 3, Systems Architecture and Design [Faisandier3 2013]) between the components when defining the architecture (maximum cohesion in each component, few interfaces between the components); an example of an architectural characteristic that goes in the direction of controlling static complexity is **modularity**. At the time of integration, care will be taken to add new components to an aggregate, only after verifying and validating each of their interfaces. For example:

Interfacing technology	Verifications to be performed
mechanics	fit, tolerance, clearance
electronics	signals, chronograms
software	syntax, semantic, scenario
communication	signals, protocols
heterogeneous (ex: mechatronics)	conversion protocol, measures

223

Controlling **dynamic complexity** would require verifying all possible scenarios (combining all possible behaviours of components and their interactions). Because of the combinatorial explosion, it is almost impossible to perform exhaustive tests, except in systems with a high security level. A certain level of confidence in integration will therefore be sought by independently validating the behaviour of each interface and defining an inspection and testing strategy leading to a reasonable compromise between their number and coverage - see section 5.1.5.

It is therefore necessary to anticipate the possibility of the existence or emergence of coupling effects. For example:

- ♦ modification of the behaviour or properties of a material component by thermal effect linked to the proximity of another heating component,
- ♦ modification of electromagnetism by the proximity of strong, weak currents or resonance,
- ♦ inter-blocking case in software and/or pollution of computer memories.

6.3.2 Notion of Integration-Enabling-System

Integration engineering and associated means can be considered as an **enabling system** that supports the operational system (System-Of-Interest, SOI) to be integrated. We call this system the Integration Enabling System (IES). The notion of enabling system is largely explained in Volume 1, System Notion and Engineering of Systems [Faisandier1 2015]. An enabling system has its own life cycle with generic states / stages and uses generic life cycle processes. The Integration Enabling System (IES) is a stand-alone system that can support the integration of one or several more or less similar systems.

The purpose of the IES is to provide a framework for preparing and executing the integration of systems of the same type. It can be seen as an infrastructure including the services, means and tools necessary to perform system integration. In its relations with a SOI to be integrated, the role of the IES is to identify, realize or obtain, then use and maintain all the resources, means and tools necessary for the integration of this SOI.

The life cycle activities of the IES interact with those of the SOI to be integrated as shown in Figure 82.

The advantage of such a provision (i.e. the separation between the SOI to be integrated and the IES) lies in the time saving by parallelising the tasks of two separate projects. It is known that the development of such an IES can take a long time, sometimes more than the development of the SOI to be integrated.

An IES can be implemented in different ways; it can be, for example, an external company, or a project separate from the SOI development to integrate. The generic IES must be instantiated to take into account the particularities of each SOI to be integrated.

Figure 82 shows some technical processes of the life cycle, such that they can be instantiated and used concurrently for the SOI to be integrated and for the IES, showing their main exchanges and how they coordinate. In this figure, the arrows between the processes (boxes) represent some of the exchanged input-output flows; these arrows do not represent the execution sequence; iterations are not represented.

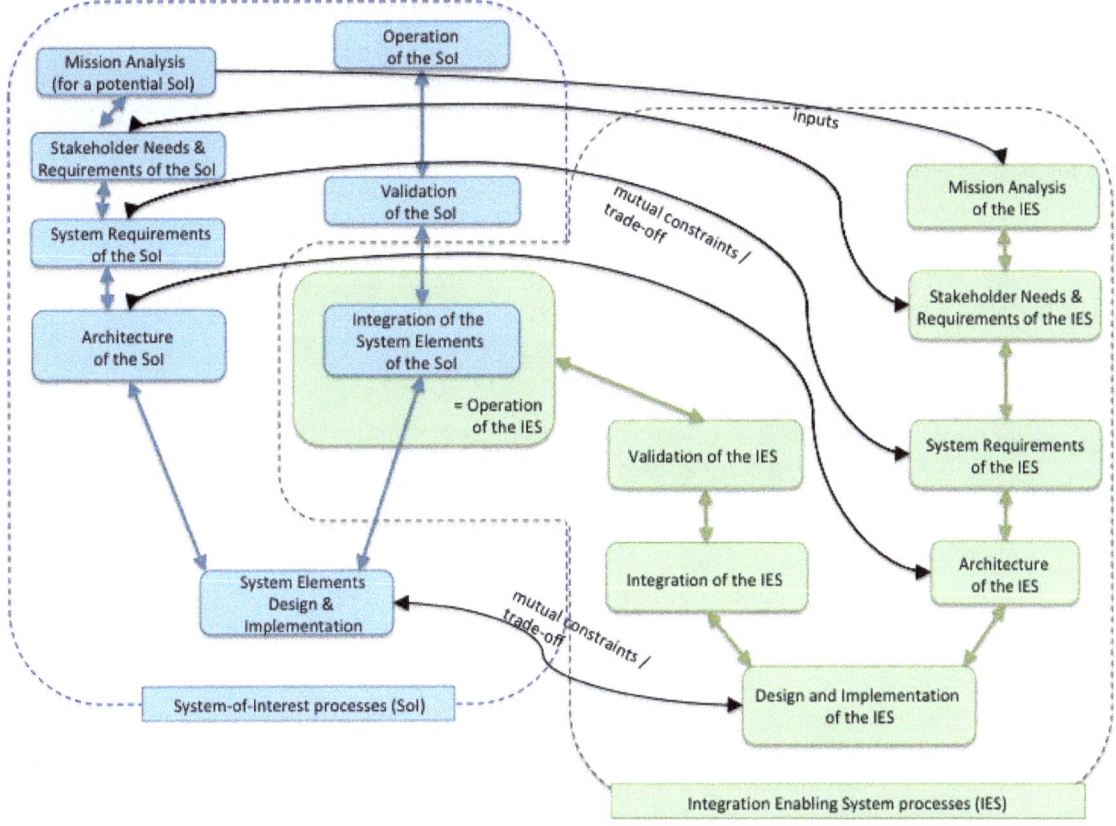

Figure 82 - Interactions between the processes used for the SOI and the IES

The life cycle technical processes applied and instantiated to the IES, which are shown in Figure 82, are briefly described below.

- The **business and/or mission analysis process** for the IES includes the preliminary study of alternative integration strategies in relation to the inputs of the operational concept and scenarios of the SOI to be integrated. This study consists of:

  ♦ analyse the operational concept and scenarios of the SOI to be integrated in order to identify and imagine the integration concepts, principles and necessary means to be put in place;

  ♦ analyse the feasibility, effectiveness, cost and risks concerning the integration of the SOI.

- The **stakeholder needs and requirements definition process** for the IES includes, in particular, identifying the stakeholders of the integration of the SOI, their expectations and constraints, and defining high-level objectives in terms of efficiency, delivery and cost. The resulting stakeholder requirements for the integration means (tools, simulators, integration procedures, verification procedures, etc.), services and their use are derived from the integration strategy of the SOI.

- The **system requirements definition process** for the IES includes defining the requirements applicable to the integration means (tools, simulators, integration procedures, verification procedures, etc.), services and their use, refining objectives (to define effectiveness requirements) and refining constraints (to define delivery times and cost limits).

- The **logical and physical architecture definition processes** for the IES include:
    - participation in refining the integration strategy of the SOI (mainly concerning the relevance of the definition of integration aggregates, verification actions and applicable techniques or methods);
    - development of a logical view of the IES; i.e., how the assembly and verification tasks are to be performed and synchronized: planning of the delivery and assembly of the components of the SOI, planning of the execution of verification actions;
    - projection / allocation of the logical view on a physical view, including adequate means (tools, simulators, integration procedures, verification procedures, etc.), resources (skills, operators, etc.);
    - the initialization of the acquisition / development of these various means via the definition of their respective requirements.
- The **design definition process** and the **implementation process** for the IES include the definition of the design characteristics and their implementation (development or reuse or manufacture or purchase of COTS) of all means (tools, simulators, integration procedures, verification procedures, etc.), as well as the training of human resources to provide qualified and experienced operators in the execution of integration procedures, verification procedures, etc.
- The **integration process** for the IES includes the acceptance and delivery of all means (tools, simulators, integration procedures, verification procedures, etc.), their assembly or connection and the verification of their interfaces and functionality.
- The **validation process** of the IES includes the validation of these means (tools, simulators, integration procedures, verification procedures, etc.) as a ready-to-operate package.
- The **operation process** of the IES is nothing other than the execution of the integration of the components of the SOI, using the means defined (tools, simulators, integration procedures, verification procedures, etc.), implemented, integrated and validated previously.

6.3.3 Interfacing with other development processes

The integration of the operational system (System-Of-Interest, SOI) must be coordinated with the other activities of the life cycle. Future integration should be considered as early as possible, which means that each life-cycle technical process should include integration-related provisions or activities. The main technical processes and provisions are as follows.

Business and/or mission analysis process

- Analyse the **concepts of operations** (businesses, markets) of the concerned organisation and the main options of any potential SOI (which would respond to the raised issue or problem, or to a business opportunity), in order to identify and imagine the corresponding integration approaches, principles and necessary means to be put in place.
- Perform feasibility, efficiency, cost and risk analyses concerning the integration approach of the potential SOI.

Stakeholder needs and requirements definition process

- Analyse **operational concepts**, operational scenarios of the SOI, associated integration approaches and principles, in order to identify and define integration needs and constraints - see section 6.1.2.

- Incorporate into the **stakeholder requirements** database the integration needs and constraints applicable to the SOI, in the form of integration type stakeholder requirements.

- From the definition of the integration strategy and following the feasibility analysis of the respect of the integration stakeholder requirements applicable to the SOI, provide feedback and recommendations about the system *integrability* (i.e. ability of the SOI to be integrated).

- When errors regarding stakeholder requirements have been identified during the execution of the integration process, the stakeholder needs and requirements definition process corrects these requirements.

System requirements definition process

- Incorporate into the **system requirements** database the integration type requirements (constraints) applicable to the SOI.

- From the definition of the integration strategy and following the feasibility analysis of the respect of the system integration requirements applicable to the SOI, provide feedback and recommendations about the system *integrability*.

- When system requirements errors have been identified during the integration process, the system requirements definition process corrects these requirements.

Logical and physical architecture definition processes

- From the definition of the integration strategy and following the analysis of the physical and logical models of the SOI, provide feedback and recommendations in order to facilitate the future integration (in particular with regard to interfaces) - see section 6.1.2.

- Architectural characteristics, such as modularity facilitate integration - see section 6.1.3.4.

- Study and identify the potential **emerging properties** and **behaviours** (intended or unintended) that may appear during the integration of components. This should be done to avoid late detection, in particular, of threats or hazards that could be generated by the system due to interactions between components.

- The **definition of interfaces** is strongly concerned by integration; the physical aspects of the interfaces (physical connection) are concerned by the assembly of the components; the logical aspects of the interfaces (input-output flows and interactions) are concerned by the execution of the functions of the integration aggregates - see section 6.1.1.

- The integration of the SOI in its context of use (existing environment) constrains the definition of interfaces.

- When architecture errors / defects (especially about interfaces) have been identified during the execution of the integration process, the logical and physical architecture definition processes correct the concerned architecture elements.

Design definition process

- The **detailed definition of the interfaces** is an important contribution to the integration strategy (in particular the order in which the components are assembled). The use of certain technologies can lead to complication of the interfaces between the components and their design, in order to allow the exchange of matter, energy and/or information (for example, by adding intermediate transformations, protocols). This may affect the order in which components are connected or assembled and/or the choice of integration methods or techniques.

- For integration purposes, **checkpoints** must be added to certain components, or their interfaces; these must be incorporated into the design definition. Moreover, checkpoints participate in the **testability** of the SOI when in service / operation.

- Virtual environments or tools (mechanical, electrical, electronic, software, operator procedures, etc.) are increasingly used when designing systems and components. They can also be useful in preparing for future integration by simulating connections of designed components, comparing aggregate definition alternatives and simulating assembly procedures. The results of these types of activities can influence the design of the system and its components - see section 6.1.3.5.

- When design errors / defects (especially about interfaces) have been identified during the integration process, the design definition process corrects the concerned design elements.

System properties assessment process (System Analysis Process)

- Perform an *integrability* analysis of predefined components (and COTS) of the SOI architecture (ability to integrate a selected component); the analysis may include an evaluation of **interoperability**.

- Perform an assessment of integration strategy alternatives to improve efficiency, reduce integration time and cost.

- Perform an assessment of verification strategy alternatives, from an integration perspective, to reduce the risks, time and cost of integration.

Verification process

- From the verification strategy, provide inputs to develop the integration strategy - see section 6.1.4.

- Perform verification actions related to the **interfaces** between the components that make up each aggregate, using selected verification techniques (e.g. inspection, tests).

- Perform verification actions to check **compliance** with architectural characteristics or design properties relevant to each aggregate.

- When errors / defects are identified, the verification process reports to the concerned processes.

Validation process

- From the validation strategy, provide inputs to develop the integration strategy, in particular data that correspond with properties that can be examined in the operational environment, such as safety and security - see section 6.1.4.

- Perform validation actions to check **compliance** with system requirements relevant to each aggregate.

- When errors / defects are identified, the validation process reports to the concerned processes.

6.3.4 Searching for and clearing defects

Within the framework of integration, the objectives of verification and validation are not only to establish the conformance of a system-of-interest (product, service or organization) with its definition references (requirements, architecture, design), but also to search for residual defects in the system-of-interest. Defect searching activity in the target product has only been mentioned a few times in this book, but has not yet been described and explained. The reason is that this activity involves most of the technical processes of the system life cycle, their articulation within development, and the corresponding skills and actors.

The search for defects in the target product / service / organization follows the observation of an anomaly during the execution of verification and validation actions - see Figure 66. The activity of searching for and clearing a defect includes the following tasks - see illustration Figure 83:

- Highlight an error
- Locate the corresponding defect
- Clear the defect (correct or repair)
- Check non-regression

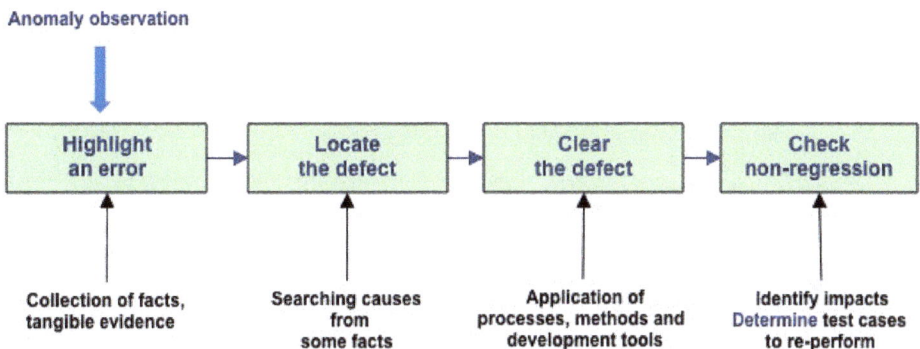

Figure 83 - Search for and clear defect

For a better understanding of this section, the reader is invited to refer to the text in section 3.5.3, *Origin of system failures*. As a reminder, a **defect** resides in a physical component for various reasons; a component performs one or more functions that can activate this defect, which causes an **error**; i.e. a result of execution different from the expected result (or absence of result). The error is therefore functional in nature and may concern the function itself (non-execution, partial execution of the required transformation, execution out of time, etc.) or the output flows (non-production of outputs, insufficient number, etc.). Here, it will be a question of highlighting an error to try to go up towards the defect.

Highlight an error

Following the finding of an anomaly attributed to a dysfunction of the system-of-interest (product / service / organization), highlighting an error requires the accumulation of facts, observations and tangible evidences on the context of the test, the conditions under which the anomaly occurred (operator actions, system reaction, messages sent, etc.). The work is greatly facilitated if the developers have been able to introduce observation elements (probes, sensors, plotters, spies, checkpoints and the corresponding observable measures) into the system. It can be useful to reproduce the anomaly on the system by performing functional type tests (by choosing the same execution conditions), then structural grey box type tests, in order to identify potentially misleading functional parts of the system, because we must not forget that an error can be caused by another error through propagation effect.

If it is not possible to use the system-of-interest due to lack of observable elements, use a replica of the system, simulators or mock-ups. Hence the importance of having such tools, which are also very useful in maintenance throughout the life of the system.

Locate the defect

This is the most complicated task if you don't have a method. Indeed, it is a question of going up the chain of causes, starting from some facts or observations. It is essentially a mental activity that looks like any type of investigation; in other words, based on facts and observations, it is a matter of formulating successive hypotheses and showing by reasoning that they are relevant or not.

The existence of functional models and behavioural models greatly facilitates the task, since these models explain how the system works, the flows of functions, the transformations from input flows to output flows.

The problem must first be clearly stated, then solved by combining symptoms and confronting them with hypotheses using different types of reasoning. Here are two useful practices concerning reasoning:

- If we block on reasoning for too long without succeeding, we make **mental fixation**. It is then better to take some time and step back by mentally disconnecting from the problem, so as to take it up again later under a new angle.

- If reasoning by a single person does not seem sufficient regarding the complexity of the problem, do not hesitate to use psychological springs in seeking listeners to whom the problem and reasoning are orally exposed. This is stimulating and also makes it possible to check the accuracy of the reasoning.

In terms of error or defect searching, beginners tend to want to use tools and experiments by trying modifications on an ad hoc basis (when possible, for example in the case of software) and observe the results. This is totally unproductive and can even be dangerous. It may also introduce new errors that hide initial errors and confuse leads.

The recommended methods are, in order of effectiveness: induction reasoning, deduction reasoning, reverse execution, step-by-step, and finally testing.

Induction reasoning

It is the method, called "from individual to general", which, in our case, consists of:

1. **Locate relevant information** based on known symptoms (facts, observations, evidences). What is the system doing or not doing correctly?

2. **Organize the facts** using the following questions: What symptoms? Where do these facts come from in the system (from what functions)? When were they produced? What is their scope, utility or importance in the system? Are there any contradictions between the facts?

3. **Study the relationships** between these facts or symptoms and try to understand the logic of the sequence of symptoms observed.

4. **Formulate hypotheses**, i.e. possible causes (defects) of these symptoms. If hypotheses cannot be made, additional clues or facts must be collected. They are obtained by performing similar test cases, but with different test data sets that do not produce these symptoms. When these facts are obtained, repeat from step 1.

5. **Prove each hypothesis** (cause) in relation to all symptoms. If demonstration is not possible, abandon this hypothesis and proceed to the next one. In the opposite case, the cause (the defect) is found, or there is a possible cause of higher level. Establish at this time a tree of causes, which will be most useful to consider one or more corrections or repairs as relevant as possible. This step is fundamental if we want to have solved the whole problem. Proof can be obtained by performing a provisional correction and mental or manual execution or using a mock-up.

Deduction reasoning

It is the so-called "general to individual" method, which, in our case, consists of:

1. **List possible causes** (defects); they do not need to be fully explained and logical at the beginning, because they are hypotheses that will serve to structure and analyse the facts or symptoms that result: production of an error, propagation of this error.
2. **Eliminate all hypotheses one by one** by analysing the symptoms looking for contradictions (same questions as in step 2 of the induction method). If all hypotheses are eliminated, other causes must be imagined and new information is usually needed.
3. **Refine defect hypotheses**. The possible cause may be exact, but not precise enough to just point out the defect. Use additional clues to obtain a more specific **theory**.
4. **Prove each hypothesis**. Refer to step 5 of the induction method.

Reverse execution

This method consists of going back up the logic of the treatments or transformations (functions) starting from the point where the results are incorrect, until the logic goes astray by observing the measures / values of the observable elements and what they should have been.

Automatic plotters or measures records can be used via **test management tools**. The availability of behavioural models is essential in order to be able to go back up the logic and compare with the intermediate results obtained on the target system.

Step by step

This method is reserved for software systems, or software parts of complex systems. Step by step is generally a facility, or software tool, offered by test means or **test management tools**. However, the method is only used to locate errors and defects in logic or calculation, except in real time.

Testing

The method for searching error or defect by testing can be used. Variants of the original test case that revealed the error should be made in order to more accurately point out the error. These variants must have execution conditions very close to those of the original test case, or else only one condition varies.

This method is used in conjunction with induction and deduction methods.

Clear the defect

Decision to correct / repair

Two essential points must be considered to correct the defect or to repair it:

1. Choose one of the two alternatives, correct / repair, or not. The alternative not to correct / repair can be chosen from the following conditions taken individually, or in combination:
   a. if the defect does not affect a function essential to the mission of the system, or to safety or security capabilities,
   b. if the defect is very complicated to correct / repair and generates costs too high for the project,
   c. if it is possible to find a solution to the problem by carrying the problem into operational constraint, for example by modifying operator procedures.
2. Implement a correction / repair. The choice of correction / repair is guided by project criteria according to the impact of this correction / repair on the technical aspects, quality, cost, deadlines. For example:
   a. a modification which is technically a bit complicated, but which simplifies the operational modes / scenarios (operator procedures, use),
   b. conversely, a simpler modification, but which complicates a little the operational modes / scenarios,
   c. a modification that impacts the minimum number of components and interfaces,
   d. a modification that does not affect the architecture, but on the contrary, the design, or the reverse,
   e. a modification that reduces effectiveness or not,
   f. a modification that requires re-running many test cases or not, or creating new test cases,
   g. etc.

Implementation of the correction / repair

The following principles must be observed in order to control residual errors and defects in the system:

- The correction / repair is carried out under the conditions of the initial development; at least under the closest conditions in terms of methods, formalisms, verifications, skills, configuration management, compliance with quality control provisions.
- Be careful not to hide or cancel only the symptoms, but correct / repair the cause(s) of the error.
- Correction or repair actions are more prone to defects than the definition and realization of the original product or service. Therefore, check the corrected or repaired parts carefully.
- One defect can hide another; it is recommended to examine the surroundings.
- A defect corrected / repaired may create a new defect elsewhere; beware of side effects by doing an impact analysis before making any changes.

Verify non-regression

Following modification, the objective is to maintain or improve the level of quality previously achieved. By quality, we mean all the characteristics that impact the satisfaction of the acquirer or users. This involves performing verifications to establish the **non-regression** of the target product, service or organization.

The two major questions concern the test cases to be carried out to obtain proof of non-regression, while having a minimum impact on costs and delays: Which test cases must be re-performed? What new test cases need to be defined and performed? The principles are as follows:

- Re-execute the test cases that revealed the anomaly.
- Re-execute all test cases relating to the elements of the repository affected by the modification, as well as the elements of the corresponding downstream repositories (cascade tests).

Figure 84 illustrates these principles.

If the modification affects the **implementation of a component** or one of its parts: re-perform the test cases of the modified parts (constituents) and the complete component to check the non-regression of functionalities, effectiveness and interfaces.

If the modification affects part of the **design and/or physical architecture**, re-perform the test cases of the modified parts, and the test cases related to the components and their interfaces impacted by this modification.

If the modification affects part of the **logical architecture**, re-perform the test cases of the modified parts, and the test cases for the impacted functions in each of the concerned components, and the integration test cases for the functions that interface.

If the modification affects the **system requirements**, re-perform the test cases of the modified parts, and the test cases of the impacted component, the integration test cases for the functions that interface and the validation test cases related to the concerned requirements.

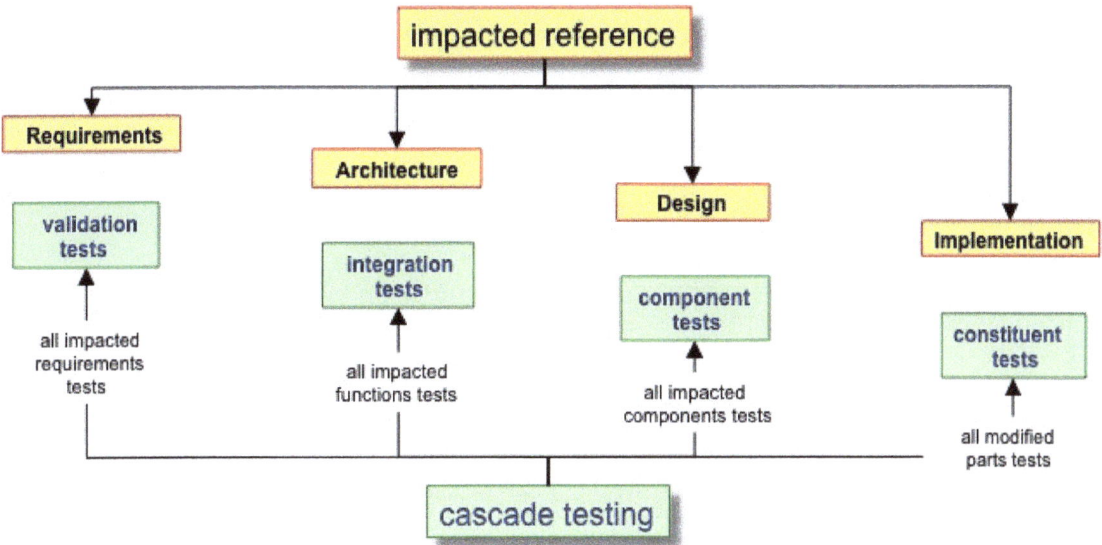

Figure 84 - Test cases to re-perform to prove non-regression

Who does what?

The tasks of searching for and correcting defects are logically divided according to the technical processes involved. Nevertheless, strong collaboration and rapid and constructive exchanges are necessary for successful and effective integration.

To summarize:

- **Error highlighting** tasks are performed by the personnel in charge of test execution and integration.

- **Defect location** tasks are carried out by personnel in charge of system definition (architects, designers) and/or implementation, supported by personnel in charge of testing and integration.

- The tasks of **defect clearing** (correction / repair) are carried out by the personnel in charge of the definition and the implementation of the system (architects, designers, trade producers).

- The **non-regression verification** tasks are performed by the personnel in charge of testing and integration, supported by the personnel in charge of system definition for the test cases definition part.

6.3.5 Recommendations and frequently asked questions

The major recommendations are as follows:

1. **Avoid the global integration method,** otherwise the detection of errors or potential defects will be long and tedious. On the contrary, it is recommended to verify the interfaces gradually by forming successive controllable aggregates. Use the coupling matrices to define the assembly order.

2. **Start integration and verification-validation preparation activities as early as possible** in the development cycle; as soon as the first assumptions on the operational concepts of the future system are imagined. Just ask the question: how will this be integrated, verified and validated?

Chapter 6.3. Integration: Practice

FAQ

Are integration and production different expressions of the same activity?

No, system integration is part of the effort devoted to making prototypes. The integration activity must not be understood as the assembly of end products on a production line.

For mass production, an assembly line uses a different assembly sequence than the integration. Integration composes (sub) systems in order to validate them almost separately (system levels).

Mass production is hardly interested in (sub) systems, but on the contrary, in sets of components to optimize assembly time and effort.

Components do not arrive in the order foreseen in the initial integration plan. What should be done to avoid delivery delays?

Indeed, experience shows that components are almost never delivered in the intended order, nor can tests be performed in the intended order.

The only solution is to provide a flexible integration strategy and verification - validation strategy; i.e. identify critical paths in a strategy and provide alternative assembly scenarios in order to reduce the overall delay and costs generated by late deliveries.

Other uses of the term integration

The term integration is used in the industrial world for other subjects or activities than that covered in this book. For example:

Integrated development: this essentially involves taking into account, during the development (mainly architecture and design), the characteristics of the product throughout its life cycle, as well as the concerned actors. It can include integrated multidisciplinary teamwork.

Models integration: These are modelling activities during development (mainly architecture and design) that consist of combining (integrating) models of the same type, or models related to several system properties; or combining product models and manufacturing models. This last activity mainly concerns design (mechanical for example).

Integrated team development: Bringing together, within a development team, the skills required to create / design and produce products or services that include several disciplines. The integrated team can be located on a single geographical site or distributed over several sites. Inter-disciplinary communication is essential for the success of the project. One of the means of implementation is complex systems engineering, seen as inter-disciplinary, as exhibited in the volumes of this series of books.

What is the difference between "measure of effectiveness (MOE)" and "measure of performance (MOP)"?

A Measure of Effectiveness is an expected effectiveness, a performance target.

A Measure of Performance is a measured or checked effectiveness on an implemented component.

7 SOME REFERENCES

The following books and documents can provide the reader with additional understandings, details or syntheses. Some following books are cited in the present volume.

Books in English:

a. [Faisandier1 2015] Faisandier Alain 2015. System Notion and Engineering of Systems. Belberaud, France : Sinergy'Com. www.sinergycom.net

b. [Faisandier2 2013] Faisandier Alain 2013. Systems Opportunities and Requirements. Belberaud, France : Sinergy'Com. www.sinergycom.net

c. [Faisandier3 2013] Faisandier Alain 2013. Systems Architecture and Design. Belberaud, France : Sinergy'Com. www.sinergycom.net

d. [Miller 1956] George A. Miller. The Magical Number Seven, Plus or Minus Two: Some Limits on our Capacity for Processing Information [archive]. The Psychological Review, 1956, vol. 63, Issue 2, pp. 81-97.

Books in French:

e. [Faisandier4 2014] Faisandier Alain 2014. Notions de système et d'ingénierie de système. Belberaud, France : Sinergy'Com. www.sinergycom.net

f. [Faisandier5 2017] Faisandier Alain 2017. Evaluation et preuve du système. Belberaud, France : Sinergy'Com. www.sinergycom.net

g. [Meinadier 1998] Meinadier Jean-Pierre. 1998. Ingénierie et intégration des systèmes. Paris: Hermes.

h. [Meinadier 2002] Meinadier Jean-Pierre. 2002. Le métier d'intégration des systèmes. Paris: Hermes-Lavoisier.

Standards, manuals, body of knowledge:

i. [INCOSE 2015] INCOSE Systems Engineering Handbook: A Guide for System Life Cycle Processes and Activities. 2015. Fourth edition. WILEY. San Diego, CA, USA: International Council on Systems Engineering (INCOSE).

j. [ISO 15288] Systems and software engineering - system life cycle processes. Geneva, Switzerland: International Organization for Standardization (ISO) / International Electronical Commission (IEC), ISO/IEC 15288:2015

k. [ISO 24748-1] Systems and software engineering - Life cycle management - Part 1: Guide for life cycle management. Geneva, Switzerland: International Organization for Standardization (ISO) / International Electronical Commission (IEC), ISO/IEC 24748-1:2016.

l. [ISO 24748-2] Systems and software engineering - Life cycle management - Part 2: Guide to the application of ISO/IEC 15288:2015. Geneva, Switzerland: International Organization for Standardization (ISO) / International Electronical Commission (IEC), ISO/IEC 24748-2:2016.

m. [ISO 24748-6] Systems and software engineering - Life cycle management - Part 6: System integration engineering. Geneva, Switzerland: International Organization for Standardization (ISO) / International Electronical Commission (IEC), ISO/IEC 24748-6:2016.

n. [NASA 2007] NASA. 2007. Systems engineering handbook. Washington, D.C.: National Aeronautics and Space Administration (NASA), NASA/SP-2007-6105.

Chapter 7. References

Web sites:

o. [SEBoK 2016] Guide to the Systems Engineering Body of Knowledge - version 1.6 - 2016 or future version. http://www.sebokwiki.org

8 ANNEXES

8.1 System Justification Document - Template and guidelines

Preliminary

This document is a form to use and fill in order to produce the System Justification Document related to a specific system in the context of a development project.

The "System Justification Document" (SysJD) presents the justification and rationale for selection of main engineering elements, integration elements (aggregates), and related verification - validation actions. Main engineering elements include the set of Stakeholder Requirements, the set of System Requirements, the architectural and designed elements: Functions, Input-output Flows, Operational Modes, Transition of Modes, Scenarios, System Elements (Components), Physical Interfaces (links / connectors).

It provides the traceability between engineering elements, and the outcomes of cost analysis, effectiveness analysis, risk analysis, dependability analysis, safety analysis, and trade-offs analysis (refer to System Analysis Process) that argue chosen engineering elements.

Chapter 8.1. Annexe: Justification Document

1 INTRODUCTION

1.1 Presentation of the document

Present briefly the document and the related concerned system. Typical content (but instantiated) could be:

This document contains elements for justification of selections and for technical decisions related to the development of System XX.

Note: The System Justification Document is filled throughout the development in order to record the justification and rationale of selections and technical decisions related to:

- *The engineering of the system to show:*
    - *relevance and consistency of the set of Stakeholder Requirements;*
    - *relevance and consistency of the set of System Requirements compared to Stakeholder Requirements (traceability tables);*
    - *relevance of the selected independent logical architecture compared to those rejected;*
    - *consistency of the selected independent logical architecture against System Requirements (traceability tables);*
    - *relevance of the selected physical architecture compared to those rejected;*
    - *consistency of the selected physical architecture against System Requirements and to logical architecture (traceability tables);*
- *The integration of the system to show:*
    - *relevance and consistency of selected verification actions against the verification objectives and constraints of the system and its systems and System Elements;*
    - *relevance of aggregates (integration sets) that justify the integration strategy.*

1.2 Overview of the system

Provide the high and synthetic description of the system, and/or refer to other documents that present it.

1.3 Documents

The documents referred here shall be defined precisely: complete title, reference and version.

They have to be placed at the disposal of the readers, either directly in annexes, or by the means of an appropriate documentation management.

The definition and distinction between 'reference documents' and 'applicable documents' must follow the rules of the company or organisation. The following two sections are suggestions for definition and distinction.

1.3.1 Reference documents

Provide the list of documents having been used for writing this document: Stakeholder Requirements Document, System Requirements Document, System Architecture Description Document, System Design Document, Justification Matrix, reports, minutes of meeting, others.

Reference	Document title

1.3.2 Applicable documents

Provide the list of documents entirely or partially applicable to the elaboration of the System Justification Document form: standards, templates, descriptions of procedure, others.

Reference	Document title

1.4 Terminology: definitions and abbreviations

Provide the list of terms and their definition used in this document absent from usual dictionaries or used with a different significance from their usual significance.

Each definition can be supplemented by the abbreviation used in the document.

Term	Definition

Chapter 8.1. Annexe: Justification Document

2 JUSTIFICATION OF SELECTED STAKEHOLDER REQUIREMENTS

2.1 Stakeholders

Provide the justification that every stakeholder has been identified and consulted.

2.2 Selection of Stakeholder Requirements

This section contains arguments / justifications that justify (when relevant only):

- *the perception of the meaning of the Stakeholder Requirement,*
- *the assessment / estimation of technical feasibility of the Stakeholder Requirement,*
- *the resolution of conflicts between any Stakeholder Requirements,*
- *the relevance of any Stakeholder Requirements related to economical, social, technical context and/or other aspects.*

3 JUSTIFICATION OF SELECTED SYSTEM REQUIREMENTS

3.1 Selection of System Requirements

This section contains arguments / justifications that justify (when relevant only):

- *the relevance of specific System Requirements, in particular effectiveness / performance requirements,*
- *the resolution of conflicts between any System Requirements (constraints, regulation documents, safety regulations, others.)*

3.2 Traceability between Stakeholder Requirements and System Requirements

This section contains the demonstration that at least one System Requirement corresponds to each Stakeholder Requirement.

The first table shows Stakeholder Requirements and correspondent System Requirements.

The second table shows the inverse in order to identify System Requirements that have been incorporated for other reasons than identified Stakeholder Requirements.

3.2.1 Correspondence Stakeholder Requirements → System Requirements

Example of table:

Stakeholder Requirement	System Requirements

3.2.2 Correspondence System Requirements → Stakeholder Requirements

Example of table:

System Requirement	Stakeholder Requirements

4 JUSTIFICATION OF LOGICAL ARCHITECTURE

4.1 Rationale for selection of the Logical Architecture

This section contains arguments / rationale that justify the relevance of the selection of the independent logical architecture compared to other candidate logical architectures:

- *Provide the description of the used decision model: assessment criteria, relative weight of criteria, and selection rules.*

- *As necessary, explain or refer to functional, behavioural and temporal models that have been used.*

- *As necessary, provide summary of simulations resulting from behavioural models that have been used to justify values of parameters or input-output flows of the behavioural architecture.*

- *Provide explanation of the pros and cons of each candidate logical architecture compared to the assessment criteria and the selection rules.*

4.2 Traceability between System Requirements and logical elements

This section provides traceability elements:

- *Each functional requirement and each functional interface requirement points at least one function of the selected logical architecture (first table).*

- *Each derived functional requirement from architecture and design and each derived functional interface requirement from architecture and design points at least one function of the selected logical architecture (first table).*

- *Each function is deduced from a System Requirement (functional, interface, operational types), or derived from the selected logical architecture (second table).*

4.2.1 Correspondence System Requirements → Functions, Operational Modes, Scenarios, Frequencies

Example of table:

System Requirement	Functions, Operational Modes, Scenarios, Frequencies

4.2.2 Correspondence Functions, Operational Modes, Scenarios, Frequencies → System Requirements

Example of table:

Function, Operational Mode, Scenario, Frequency	System Requirements

5 JUSTIFICATION OF PHYSICAL ARCHITECTURE

5.1 Rationale for selection of the Physical Architecture

This section contains arguments / rationale that justify the relevance of the selection of the physical architecture compared to other candidate physical architectures:

- *Provide the description of the used decision model: assessment criteria, relative weight of criteria, and selection rules.*
- *As necessary, explain or refer to physical architecture models that have been used.*
- *Provide explanation of the pros and cons of each candidate physical architecture compared to assessment criteria and selection rules.*

This section also contains (reused or new) heuristics for allocation of Operational Modes, Scenarios, Functions, and Input-output Flows of the selected logical architecture to System Elements, technologies, and physical interfaces (links / connectors) of the physical architecture.

It contains arguments / rationale that have been used to solve conflicts between System Elements and technologies of the selected physical architecture.

Example of table:

Rationale	Description	System Element or Physical Interface	Reference file

5.2 Traceability between System Elements, Functions and System Requirements

This section provides traceability elements:

- *Each Function of the logical architecture is performed at least by one System Element of the physical architecture (first table).*
- *Each Input-output Flow of the logical architecture is carried by one physical interface (link / connector) of the physical architecture (second table).*
- *System Requirements points System Elements and associated Design Properties (third table).*
- *Each System Element performs at least one Function (fourth table).*

- *Each link / connector carries one Input-output Flow (fifth table).*
- *Each System Element points any direct or derived System Requirements (sixth table).*

5.2.1 Correspondence Functions → System Elements

Example of table:

Function	System Element

5.2.2 Correspondence Input-output Flows → Physical Interfaces

Example of table:

Input-output Flow, trigger	Physical Interface (link / connector)

5.2.3 Correspondence System Requirements → System Elements

Example of table:

System Requirement	System Elements	Concerned Design Properties

5.2.4 Correspondence System Elements → Functions

Example of table:

System Element	Functions

5.2.5 Correspondence Physical Interfaces → Input-output Flows

Example of table:

Physical Interface (link / connector)	Input-output Flow, trigger

5.2.6 Correspondence System Elements → System Requirements

Example of table:

System Element	System Requirements

6 JUSTIFICATION OF VERIFICATION AND VALIDATION ACTIONS

Provide the major arguments / justifications for the choice of verification and validation actions in relation to the objectives (coverage rates) and the verification and validation constraints of the system, see section 5.1.5 paragraph Selection of verification - validation actions.

Refer to the Justification Matrix, see Annexe 8.3.

7 JUSTIFICATION OF INTEGRATION AGGREGATES

Provide the major arguments / justifications or approach for the choice of aggregates that justify the integration strategy, see section 6.1.3.

8 TECHNICAL RISKS MITIGATION

Provide the list of technical risks generated by engineering elements (Stakeholder Requirement, System Requirement, Function, System Element, etc.) that have been studied, mitigated, and possibly still open.

Provide the list of technical risks generated by the choice of verification and validation actions (not selected actions, selected techniques, implemented means, etc.) that have been studied, mitigated, and possibly still open.

Example of table:

Risk	Description	Causes	Status

8.2 System Verification and Validation Plan - Template and guidelines

Preliminary

This document is a form to use and fill in order to produce the System Verification and Validation Plan related to a specific system in the context of a development project.

The System Verification and Validation Plan describes the strategy for verifying and validating the system, activities and tasks for defining and executing verification - validation actions, resources, organization, deliverables, means and tools, schedules.

The System Verification and Validation Plan is consistent with the System Integration Plan and with the Project Management Plan.

Chapter 8.2. Annexe: Verification and Validation Plan

1 INTRODUCTION

1.1 Presentation of the document

Present briefly the document and the related concerned system. Typical content (but instantiated) could be:

This document describes the strategy selected to verify and validate the system XX, the necessary means and tools, the schedules, the activities and the organization.

Note: The Verification and Validation Plan is initialized as soon as the first stakeholder requirements are produced.

1.2 Overview of the system

Provide the synthetic description of the system, and/or refer to other documents.

1.3 Documents

The documents referred here shall be defined precisely: complete title, reference and version.

They have to be placed at the disposal of the readers, either directly in annexes, or by the means of an appropriate documentation management.

The definition and distinction between 'reference documents' and 'applicable documents' must follow the rules of the company or organisation. The following two sections are suggestions for definition and distinction.

1.3.1 Reference documents

Provide the list of documents used to elaborate the technical content of the Verification and Validation Plan: Stakeholder requirements document, System requirements document, System architecture description document, reports, minutes of meetings, others.

Reference	Title of the document

1.3.2 Applicable documents

Provide the list of documents entirely or partially applicable to the elaboration of the System Verification and Validation Plan form: standards, templates, descriptions of procedure, others.

Reference	Title of the document

1.4 Terminology: definitions and abbreviations

Provide the list of terms and their definition used in this document absent from usual dictionaries or used with a different significance from their usual significance.

Term	Definition

2 VERIFICATION AND VALIDATION OBJECTIVES AND CONSTRAINTS

Determine the verification and validation scope that defines the verification and validation actions to be performed. For that:

a) List the applicable system requirements, architectural characteristics and design properties, or refer to the Justification Matrix, see Annexe 8.3:

- *Stakeholder requirements*
- *System requirements*
- *Contractual requirements*
- *Normative, legislative, approval, certification requirements*
- *Architectural characteristics*
- *Design properties*

Note: if requirements or properties are not directly verifiable, they are then refined into more elementary requirements or properties that are themselves verifiable. This work is sometimes necessary to obtain satisfactory coverage of requirements and properties. This may, for example, be the case for operational requirements.

b) Identify and list constraints that limit the execution of verification and validation actions:

- *Technical constraints: criticality of the system, destructive tests, impossibility of adding check points, regulations prohibiting certain types of tests, technical unfeasibility of means or tools, etc.*
- *Project constraints: costs of verification - validation actions and associated means and tools, deadlines for the realization of associated means and tools, unavailability of means and qualified personnel, industrial organization, regulation, etc.*
- *Contractual constraints: level of reliability required and associated demonstration, organisation of tests, etc.*

3 VERIFICATION - VALIDATION STRATEGY

3.1 Adopted principles

Indicate the principles adopted concerning:

- *the selected verification - validation techniques (inspections, tests, reviews, etc.)*
- *the distribution of the execution of verification and validation actions during phases and milestones*
- *the means, tools or enabling systems to be put in place*

- the distribution of the execution of verification and validation actions between system levels, upper systems, sub-systems, etc.
- the assignment of priorities

3.2 Verification and Validation Actions

Indicate / list the selected Verification and Validation Actions that result from the comparison of objectives and constraints by applying the selected principles.

Justify the Verification and Validation Actions rejected at the time of analysis.

Indicate when the selected Verification and Validation Actions are performed over time, in what order: phases, milestones of the system development cycle.

Indicate the criteria for stopping Verification and Validation Actions and the values of these criteria. These criteria are often expressed as expected coverage rates for different types of requirements and properties.

Note: this chapter can outline and refer to the Justification Matrix, see Annexe 8.3.

4 VERIFICATION - VALIDATION ACTIVITIES

List verification and validation activities, tasks and artefacts. For this, instantiate for the system XX the activities, tasks and generic artefacts indicated in the corresponding processes in the sections 5.2.1.3 , 5.2.2.3 , 5.2.1.5 and 5.2.2.5.

5 ORGANISATION AND RESPONSIBILITIES

Indicate the structure and composition of the verification - validation team(s), respective roles.

Indicate the relationships of the verification - validation team with the other teams, as well as the assignment of tasks.

Indicate the responsibilities of each role.

6 MEANS AND TOOLS

Human resources: *list for each role, the skills required and their availability, the related training.*

Means and tools: *list for each means and tool, the acquisition mode: purchase, rental, development, subcontracting, availability (dates).*

Places where verification and validation procedures are carried out: *list for each location, infrastructures, platforms, simulators, etc.*

7 SCHEDULE

This is the schedule for the execution of the verification - validation activities and tasks listed in paragraph 4 of this plan.

For the performance of the verification - validation procedures, present a summary of the dates of the verification and validation Actions indicated in paragraph 3.2 of this plan.

8.3 Justification Matrix

The justification matrix is generally realized by a database; it can also be annexed to the verification - validation plan. It is established iteratively; its filling begins as soon as the requirements are produced; this makes it possible to validate the requirements. It is a tool to help validate compliance with requirements. It contains the complete list of system requirements and properties to be verified on the concerned system.

For each requirement and property, provide the following information:

- *the verification - validation action to be performed to establish compliance with the requirement or property*
- *the object on which the action is performed*
- *the priority (for example marked from 1 to 3)*
- *the origin, for example:*
    - *requirements documents*
    - *architecture and design description documents*
    - *agreement*
    - *legislation, standards*
    - *certification of the product*
    - *other system-block (deported verification)*
    - *etc.*

   Note: A requirement or property may have several different origins.

- *the used verification technique or method:*
    - *mathematical, logical or analogy analysis*
    - *documentary or physical inspection*
    - *virtual simulation*
    - *physical simulation*
    - *review*
    - *test*
    - *not covered (verification will not be performed)*

   Note: The same requirement or property can be verified in several different ways, in different phases or milestones.

- *the justification for not covering; indicate the argument which led to the withdrawn of the verification (for 'non-covered' requirements only); the reason may be :*
    - *a too high cost*
    - *a too long time of realization*
    - *a delay in the schedule leading to the deletion of tasks*
    - *the transfer of the verification to another system block (upstream or downstream)*

- ♦ unauthorised destruction of the product
- ♦ other
- the phase or milestone during which the verification is performed; for example :
  - ♦ expression of needs and requirements
  - ♦ definition of architecture and design
  - ♦ realization
  - ♦ integration
  - ♦ final validation
  - ♦ Requirements Review
  - ♦ Preliminary Design Review
  - ♦ Design Review
  - ♦ Validation Preparation Review
  - ♦ Validation Review
  - ♦ other
- the reference of the "verification - validation action sheet" (see Annexe 8.4)
- the reference of configuration of the object submitted to verification - validation
- the reference of configuration of the used means and tools
- the verification - validation progress status, for example:
  - ♦ defined
  - ♦ in progress
  - ♦ completed
- the synthesis of the verification - validation result:
  - ♦ positive
  - ♦ negative and reason: non conformance
- the reference of the verification - validation report

Example of justification matrix

The following example is limited to the verification - validation of FITVEE product (system realised) against its system requirements; design properties are not taken into account. An example of each requirement type is discussed, but can be generalized.

Chapter 8.3. Annexe: Justification Matrix

Ref	Requirement /property (Reference for comparison)	Verification or Validation Action (Type of action)	Concerned object	Used Technique	Phase or project activity	Execution level
Functional requirements						
Rq.F (Mission)	FITVEE system intervenes overland on forest and home fires, in coordination with other means of the district	Verification that the mission is taken into account	Functional Architecture	Inspection	Architecture	FITVEE
		Validation of the scenario "Context Functional Architecture"	System	Validation testing	Final validation	FITVEE context
Rq.F.1	FITVEE fights against forest and home fires					
Rq.F.1.1	FITVEE projects water on the fires					
Rq.F.1.2	FITVEE controls extinction of fires	Verification of respect for the rules and conventions of the Logical Architecture	Associated behavioural Scenario	Inspection	Logical Architecture	FITVEE
		Verification of respect for the rules and conventions of the Physical Architecture	Concerned sub-system(s)	Inspection	Physical Architecture	
		Validation of the matching between this requirement and the logical architecture functions	Associated behavioural Scenario	Allocation Matrix Requirements / Architecture	Architecture	
		Verification of the implementation of the Scenario for this requirement	System being integrated	Integration testing	Integration	
		Validation of the operational Scenario "Intervention on Fire Zone"	Complete system	Validation testing	Final validation	
Rq.F.2	FITVEE is deployed in the entire district					
Rq.F.2.1	FITVEE moves on paved road					
Rq.F.2.2	FITVEE moves on muddy, sandy and stony land					
Rq.F.2.3	FITVEE crosses the obstacles that separate it from fire					

Chapter 8.3. Annexe: Justification Matrix

Ref	Requirement / property (Reference for comparison)	Verification or Validation Action (Type of action)	Concerned object	Used Technique	Phase or project activity	Execution level
Functional requirements						
Rq.F.3	FITVEE reinforces and supports another FITVEE system					
Rq.F.3.1	In passive intervention, FITVEE provides water to another FITVEE					
Rq.F.3.2	In passive intervention, FITVEE provides fuel to another FITVEE	Verification of respect for the rules and conventions of the Logical Architecture	Associated behavioural Scenario	Inspection	Logical Architecture	FITVEE
		Verification of respect for the rules and conventions of the Physical Architecture	Concerned sub-system(s)	Inspection	Physical Architecture	
		Validation of the matching between this requirement and the logical architecture functions	Associated behavioural Scenario	Allocation Matrix Requirements / Architecture	Architecture	
		Verification of the implementation of the Scenario for this requirement	System being integrated	Integration testing	Integration	
		Validation of the operational Scenario " Go as reinforcement other FITVEE "	Complete system	Validation testing	Final validation	
Rq.F.3.3	FITVEE transmits demands of reinforcement to COCOCE					
Rq.F.4	FITVEE stores water and fuel					
Rq.F.4.1	FITVEE fills up water from the water distribution network					
Rq.F.4.2	FITVEE fills up fuel from the fuel distribution stations					
Rq.F.5	FITVEE exchanges information with COCOCE to coordinate its actions with other means of SSF					
Rq.F.5.1	FITVEE transmits and receives communications to/from COCOCE					
Rq.F.6	FITVEE realizes interventions of first aid to injured persons					

Chapter 8.3. Annexe: Justification Matrix

Ref	Requirement / property (Reference for comparison)	Verification or Validation Action (Type of action)	Concerned object	Used Technique	Phase or project activity	Execution level
Effectiveness requirements (performance)						
Rq.E.1	FITVEE runs over 30 km/h average on all grounds					
Rq.E.2	FITVEE runs over 90 km/h average on paved road					
Rq.E.3	FITVEE is deployed in less than 15 minutes					
Rq.E.4	FITVEE passes slopes of 30% and banking of 15%					
Rq.E.5	FITVEE delivers a mini water flow of 50 m3/h	Verification that effectiveness is taken into account	Candidate Architectures	Inspection	Logical Architecture	FITVEE
		Validation of the matching between this requirement and the logical architecture effectiveness	Candidate Architectures	Allocation Matrix Requirements / Architecture	Architecture	
		Verification of the allocation to the concerned sub-system	Requirement of the concerned sub-system	Inspection	Architecture	
		Validation of the operational Scenario "Intervention on Fire Zone"	Complete system	Validation testing	Validation	
	Requirement of the concerned sub-system	Verification that effectiveness is taken into account	Design of the concerned sub-system	Inspection	Realization	Concerned sub-system
		Validation of the matching between this requirement and the concerned sub-system design	Design of the concerned sub-system	Simulation	Realization	
		Validation of the respect of this requirement by the concerned sub-system	Product	Test	Validation	
Rq.E.6	FITVEE delivers a mini water pressure of x Pa					
Rq.E.7	FITVEE has a cruise range of minimum movement of 500 km					
Rq.E.8	FITVEE equipment has a cruise duration of at least 6 consecutive hours					

255

Chapter 8.3. Annexe: Justification Matrix

Ref	Requirement /property (Reference for comparison)	Verification or Validation Action (Type of action)	Concerned object	Used Technique	Phase or project activity	Execution level
Functional interface requirements						
Rq.IF.1	In active intervention, FITVEE receives water from another FITVEE					
Rq.IF.2	In active intervention FITVEE receives fuel from another FITVEE					
Rq.IF.3	In passive intervention, FITVEE provides water to another FITVEE					
Rq.IF.4	In passive intervention, FITVEE provides fuel to another FITVEE					
Rq.IF.5	FITVEE receives orders of intervention and information from COCOCE	Verification of respect for the rules and conventions of the Logical Architecture	Associated behavioural Scenario	Inspection	Logical Architecture	FITVEE
		Verification of respect for the rules and conventions of the Physical Architecture	Concerned sub-system(s)	Inspection	Physical Architecture	
		Validation of the matching between this requirement and the logical architecture functions	Associated logical & Physical Architecture	Allocation Matrix Requirements / Architecture	Architecture	
		Verification of the implementation of the associated Scenario	System being integrated	Integration testing	Integration	
		Validation of the operational Scenario "Intervention on Fire Zone"	Complete system	Validation testing	Final Validation	
Rq.IF.6	FITVEE receives information about actions of sprinklers planes WATERAIR from COCOCE					
Rq.IF.7	FITVEE transmits demands of WATERAIR sprinkler service to COCOCE					
Rq.IF.8	FITVEE receives information about support actions of another FITVEE from COCOCE					
Rq.IF.9	In active or passive intervention, FITVEE sends reports of the status to COCOCE					

Chapter 8.3. Annexe: Justification Matrix

Ref	Requirement / property (Reference for comparison)	Verification or Validation Action (Type of action)	Concerned object	Used Technique	Phase or project activity	Execution level
Physical interface requirements						
Rq.IP.1	Water flexible connectors of FITVEE are compatible with fire hydrants located throughout the country	Verification that this requirement is taken into account	Candidate Physical Architectures	Inspection	Physical Architecture	FITVEE
		Validation of the matching between this requirement and the physical architecture	Candidate Physical Architectures	Allocation Matrix Requirements / Architecture	Architecture	
		Verification of the allocation to the concerned sub-system	Requirement of the concerned sub-system	Inspection	Architecture	
	Requirement of the concerned sub-system	Verification that this requirement is taken into account	Design of the concerned sub-system	Inspection	Realization	Concerned sub-system
		Validation of the matching between this requirement and the concerned sub-system design	Design of the concerned sub-system	Expertise or Simulation	Realization	
		Validation of the respect of this requirement by the concerned sub-system	Product	Validation testing	Validation	
Rq.IP.2	FITVEE communicates through the mobile phone network					
Rq.IP.3	The transmission frequency of FITVEE is compatible with that of COCOCE					

Chapter 8.3. Annexe: Justification Matrix

Ref	Requirement /property (Reference for comparison)	Verification or Validation Action (Type of action)	Concerned object	Used Technique	Phase or project activity	Execution level
Operational modes requirements						
Rq.Mod.1	FITVEE decides its operational modes: In its life, FITVEE decides its operational modes according to orders received, information received, and status of its equipment. These operational modes are: awake, exercise, maintenance, active intervention (intervention on fire zone), and passive intervention (support for another FITVEE).	Verification of the consistency between the description and the sequence of the different modes	Scenario "Sequence of modes"	Inspection	Architecture	FITVEE
		Verification of the Scenario "Sequence of modes"	System being integrated	Integration testing	Integration	
		Validation in relation to the operational scenario of use	Complete system	Validation testing	Final Validation	
Rq.Mod.2	Awake Mode: FITVEE performs the following actions: perform the level-1 operation of maintenance, establish the planning of staff, and fill up with water and fuel.					
Rq.Mod.3	Active Intervention Mode: FITVEE performs the following actions: move, identify the type of fire, establish the attack strategy, deploy, fight fires, and provide first aid as necessary.	Verification of the consistency between this requirements and the functions of the logical architecture	Function "Intervene on fires"	Inspection	Logical Architecture	FITVEE
		Validation in relation to the operational scenario of use	Complete system	Validation testing	Final Validation	
Rq.Mod.4	Passive Intervention Mode: FITVEE performs the following actions: move, supply with water and fuel, and provide first aid if necessary.					
Rq.Mod.5	Exercise Mode: FITVEE performs the following actions: train to use the equipment of fire fighting, etc.					
Rq.Mod.6	Maintenance Mode: FITVEE performs the following actions: To Be Defined					

Chapter 8.3. Annexe: Justification Matrix

Ref	Requirement / property (Reference for comparison)	Verification or Validation Action (Type of action)	Concerned object	Used Technique	Phase of project activity	Execution level
Human factors requirements						
Rq.Hm.1	FITVEE is operated by 4 operators maximum	Verification that this requirement is taken into account	Operability study	Inspection	Architecture	FITVEE
		Validation of the relevance of the study conclusions	Operability study	Expertise	Architecture	
		Verification of the consistency between this requirements and the architecture	Candidate Architectures	Allocation Matrix / Requirements / Architecture	Architecture	
		Verification of the allocation to the concerned sub-systems	Requirements of the concerned sub-systems	Inspection	Architecture	
		Validation in relation to the operational scenario of use	Complete system	Validation testing	Final Validation	
	Requirements of the concerned sub-systems	Verification that these requirements are taken into account	Design of the concerned sub-systems	Inspection	Realization	Concerned sub-systems
		Validation of the matching between this requirement and the concerned sub-systems design	Design of the concerned sub-systems	Simulation	Realization	
		Validation of the respect of this requirement by the concerned sub-systems	Products	Test	Realization	
Availability requirements						
Rq.Av.1	The availability of FITVEE is higher than 70% on one year period					
Environmental requirements						
Rq.Ev.1	FITVEE is mobile in 40cm of water					
Rq.Ev.2	FITVEE is mobile in 40cm of snow					
Rq.Ev.3	FITVEE is operational in winds of 120km/h					
Rq.Ev.4	FITVEE resists heat (400° C from 20 m during 30 minutes)					

Chapter 8.3. Annexe: Justification Matrix

Ref	Requirement /property (Reference for comparison)	Verification or Validation Action (Type of action)	Concerned object	Used Technique	Phase or project activity	Execution level
Transportation requirements						
Rq.Tr.1	FITVEE vehicle is equipped with fixings for stowing on railway wagons	Verification that this requirement is taken into account	Candidate Physical Architectures	Inspection	Physical Architecture	FITVEE
		Validation of the consistency between this requirement and the architecture	Candidate Physical Architectures	Allocation Matrix Requirements / Architecture	Architecture	
		Verification of the allocation to the concerned sub-system	Requirement of the concerned sub-system	Inspection	Architecture	
	Requirement of the concerned sub-system	Verification that this requirement is taken into account	Design of the concerned sub-system	Inspection	Realization	Concerned sub-system
		Validation of the consistency between this requirement and the design of the concerned sub-system	Design of the concerned sub-system	Expertise or Simulation	Realization	
		Validation of the respect of this requirement by the concerned sub-system	Product	Validation testing	Validation	
Rq.Tr.2	Height of FITVEE vehicle is less than 3.5 m	Verification that this requirement is taken into account	Candidate Physical Architectures	Inspection	Architecture	FITVEE
		Validation of the consistency between this requirement and the architecture	Candidate Physical Architectures	Allocation Matrix Requirements / Architecture	Architecture	
		Verification of the allocation to the concerned sub-systems	Requirement of the concerned sub-system	Inspection	Architecture	
	Requirement of the concerned sub-systems	Verification that this requirement is taken into account	Design of the concerned sub-system	Inspection	Realization	Concerned sub-system
		Validation of the consistency between this requirement and the design of the concerned sub-systems	Design of the concerned sub-system	Expertise	Realization	
		Validation of the respect of this requirement by the concerned sub-systems	Products	Inspection	Realization	
Rq.Tr.3	Weight of FITVEE vehicle is lower than 38 tons					

Chapter 8.3. Annexe: Justification Matrix

Ref	Requirement / property (Reference for comparison)	Verification or Validation Action (Type of action)	Concerned object	Used Technique	Phase or project activity	Execution level
Physical constraints requirements						
Rq.Cp.1	FITVEE has a mini water tank of 20 m3	Verification that this constraint is taken into account	Candidate Physical Architectures	Inspection	Architecture	FITVEE
		Validation of the consistency between this constraint and the architecture	Candidate Physical Architectures	Allocation Matrix Requirements / Architecture	Architecture	
		Verification of the allocation to the concerned sub-system	Requirement of the concerned sub-system	Inspection	Architecture	
	Constraint of the concerned sub-system	Verification that this constraint is taken into account	Design of the concerned sub-system	Inspection	Realization	Concerned sub-system
		Validation of the consistency between this constraint and the design of the concerned sub-system	Design of the concerned sub-system	Expertise	Realization	
		Validation of the respect of this constraint by the concerned sub-system	Product	Inspection	Realization	
Rq.Cp.2	Vehicle size of FITVEE system is compatible with infrastructure of road network					

Chapter 8.3. Annexe: Justification Matrix

Ref	Requirement/property (Reference for comparison)	Verification or Validation Action (Type of action)	Concerned object	Used Technique	Phase or project activity	Execution level
Realisation constraint requirements						
Rq.Cd.1	FITVEE owns a protected and air-conditioned passenger compartment					
Rq.Cd.2	Equipment of FITVEE use non-inflammable materials					
Rq.Cd.3	FITVEE includes two heat sensors	Verification that this constraint is taken into account	Candidate Architectures	Inspection	Architecture	FITVEE
		Validation of the consistency between this constraint and the architecture	Candidate Architectures	Allocation Matrix Requirements / Architecture	Architecture	
		Verification of the allocation to the concerned sub-system	Requirement of the concerned sub-system	Inspection	Architecture	
	Constraint of the concerned sub-system	Verification that this constraint is taken into account	Design of the concerned sub-system	Inspection	Realization	Concerned sub-system
		Validation of the consistency between this constraint and the design of the concerned sub-system	Design of the concerned sub-system	Expertise or Simulation	Realization	
		Validation of the respect of this constraint by the concerned sub-system	Product	Inspection	Realization	
Rq.Cd.4	FITVEE includes two water pumps					
Rq.Cd.5	FITVEE includes two fuel pumps					
Rq.Cd.6	The rolling base of FITVEE (chassis and engine) is an existing industrial model in the range of 1000-1500 cv					
Rq.Cd.7	The water cannons of FITVEE are existing industrial models in the range of 40 L/mn flow					
Rq.Cd.8	The mechanism of heat detection in FITVEE uses infrared technology					
Rq.Cd.9	60% weight of the vehicle materials used are recyclable					

Chapter 8.3. Annexe: Justification Matrix

Ref	Requirement /property (Reference for comparison)	Verification or Validation Action (Type of action)	Concerned object	Used Technique	Phase or project activity	Execution level
Maintenance / support requirements						
Rq.Cm.1	The maintenance of Level-1 material equipment is performed in less than 30 minutes	Verification that this requirement is taken into account	Maintenance studies	Inspection	Architecture	FITVEE
		Validation of the relevance of the studies conclusions	Maintenance studies	Expertise	Architecture	
		Validation of the consistency between this requirement and the architecture	Candidate Architectures	Allocation Matrix Requirements / Architecture	Architecture	
		Verification of the allocation to the concerned sub-systems	Requirements of the concerned sub-systems	Inspection	Architecture	
		Validation in relation to the operational scenario of the Maintenance mode	Complete system	Validation testing	Final Validation	
	Requirements of the concerned sub-systems	Verification that these requirements are taken into account	Design of the concerned sub-systems	Inspection	Realization	Concerned sub-systems
		Validation of the consistency between these requirements and the design of the concerned sub-systems	Design of the concerned sub-systems	Simulation	Realization	
		Validation of the respect of these requirements by the concerned sub-systems	Products	Test	Realization	
Rq.Cm.2	The equipment of FITVEE system is repaired by a maintenance system for levels higher than 1					

Chapter 8.3. Annexe: Justification Matrix

Ref	Requirement /property (Reference for comparison)	Verification or Validation Action (Type of action)	Concerned object	Used Technique	Phase or project activity	Execution level
Regulatory requirements						
Rq.Cr.1	The vehicle of FITVEE is in standard colour for fire-fighting vehicles					
Rq.Cr.2	FITVEE is equipped with standard acoustic and illuminated fire-alarm for fire-fighting vehicles					
Rq.Cr.3	FITVEE respects the traffic regulations	Verification that the traffic regulations are taken into account	Architecture studies	Inspection	Architecture	FITVEE
		Verification that the studies conclusions are taken into account	Candidate Architectures	Inspection	Architecture	
		Validation of the consistency between this requirement and the architecture	Candidate Architectures	Allocation Matrix Requirements / Architecture	Architecture	
		Verification of the allocation to the concerned sub-system	Requirement of the concerned sub-system	Inspection	Architecture	
	Requirement of the concerned sub-system	Verification that this requirement is taken into account	Design of the concerned sub-system	Inspection	Realization	Concerned sub-system
		Validation of the consistency between this requirement and the design of the concerned sub-system	Design of the concerned sub-system	Expertise	Realization	
		Verification of the respect of this requirement by the concerned sub-system	Product	Inspection	Realization	
		Validation of the respect of this requirement by the concerned sub-system	Product	Test	Realization	
Validation requirements						
Rq.V.1	FITVEE performs two tests in operational conditions according to protocol XXX	Verification that this requirement is taken into account	Final Validation Plan	Inspection	Validation Plan development	FITVEE
		Validation of the relevance of the test	Execution of the test	Expertise	Final Validation	

8.4 Verification - Validation Action Sheet

This sheet indicates, for each verification - validation action, the following information:
- *Identifier and description of the action*
- *the subject of the action*
- *used references*
- *used technique or method*
- *development phase or activity during which the action is carried out*
- *concerned system level*
- *expected result*

8.5 Verification - Validation Procedure Sheet

Each procedure contains the following information:

1 PROCEDURE

Identifier and name or title of the procedure

2 VERIFICATION OR VALIDATION PREPARATION

Concerned verification - validation action(s): refer to the Verification - Validation Action Sheet(s)

List of means and tools required: harnesses, simulators, operator resources, enabling systems, etc.

Configuration: configuration of the object submitted for verification - validation, means and tools, actions, parameter values, necessary quantity entries, etc.

3 EXECUTION OF THE PROCEDURE

Orderly description of execution tasks that includes:

- *initial conditions*
- *step-by-step tasks*
- *external stimuli or commands to be given by operators or tools to initiate or execute verification and validation actions (possibly refer to the descriptive sheets of these verification and validation actions)*
- *intermediate check points and results to be recorded*

Description of results (after execution of the procedure) that includes:

- *list of results obtained from the execution of verification - validation actions*
- *list of encountered findings, difficulties and anomalies during the execution of the procedure*

Synthetic conclusion: OK or not OK, decisions (accept, redo, rejected, etc.)

8.6 Anomaly or Non-conformance Sheet

The verification and validation teams make observations during the execution of the verification - validation actions via the Verification - Validation Procedures. These observations are differences with respect to an expected outcome and give rise to anomaly records.

The analysis of these anomalies may give rise to non-conformances, if the finding turns out to be a difference with respect to a reference of the system definition, or to suggestions for correction / repair or improvement.

An anomaly or non-conformance sheet contains the following information:

- *Information of the context:*
    - *Product identification, version*
    - *Name and contact of issuer of the report*
    - *Description of symptoms or findings: describe what happened and what led to the anomaly, how the anomaly manifested itself (conditions), the visible consequences of the anomaly, the damage(s) caused*

- *Synthesis of the anomaly:*
    - *Short text of the anomaly*
    - *Date of occurrence*
    - *Number of occurrence*
    - *Impact of the anomaly: visible or tangible consequences, damage(s) caused by the anomaly as an observable failure / malfunction*
    - *Actors concerned*
    - *Classification of the anomaly: according to the criteria of nature, impact and repetitiveness, the anomaly is classified as minor (we can continue with the product as is), or major (the product must be repaired or corrected)*

- *Analysis of the anomaly: the goal is to go back to the causes according to the approach Failure / observable finding - Error - Defect - Hypothesis of mistake*
    - *Sequence of events that took place before the anomaly and that induce a potential defect*
    - *Decision of non-conformance or not with a system definition reference*
    - *Suggestions for corrective action (to eliminate the defect) or improvement (so that the defect is not repeated)*

8.7 Inspection Sheet

This sheet is used to exchange between the inspector and the developer. It has three information areas:

- Area 1: **Identification**
  - Name of the issuer and date
  - Identification number of the sheet
- Area 2: **Object, Verification validation Action, Reference**
  - Object: document inspected, date and reference of the document, author name
  - Verification - validation Action: characteristic(s), analysed property (ies) by the inspector
  - Reference: document(s) referred to
- Area 3: **Comments and Questions asked**
- For example in the form of a three-column table:
  - Paragraph number / document page
  - Statement of Comment / Question
  - Author's answer: OK / non OK with justification

8.8 Problem Sheet

This sheet is used to express problems or questions in writing. It contains the following information:

- Area 1: **Identification**
  - Name of the issuer and date
  - Identification number of the sheet
  - Name of the concerned review
  - Document concerned
  - Concerned product, or service or system
- Area 2: **For the Review Group use only**
  - Statement of the problem or question asked by the member of the review group
- Area 3: **For the Project Team use only**
  - Response provided by the project team
- Area 4: **For the Steering Committee use only**
  - Decision taken by the Steering Committee

8.9 System Integration Plan - Template and guidelines

Preliminary

This document is a form to use and fill in order to produce the System Integration Plan related to a specific system in the context of a development project.

The System Integration Plan describes the strategy for integrating system components, activities and tasks to perform integration engineering, resources, organization, deliverables, means and tools, schedules.

The System Integration Plan is consistent with the System Verification and Validation Plan and the Project Management Plan.

1 INTRODUCTION

1.1 Presentation of the document

Present briefly the document and the related concerned system. Typical content (but instantiated) could be:

This document describes the strategy chosen to integrate the components of System XX, the deliverables, the means and tools required, the schedules, the activities and the organization.

Note: The Integration Plan is initialized as soon as the first system architecture elements are produced.

1.2 Overview of the system

Provide the synthetic description of the system, and/or refer to other documents.

1.3 Documents

The documents referred here shall be defined precisely: complete title, reference and version.

They have to be placed at the disposal of the readers, either directly in annexes, or by the means of an appropriate documentation management.

The definition and distinction between 'reference documents' and 'applicable documents' must follow the rules of the company or organisation. The following two sections are suggestions for definition and distinction.

1.3.1 Reference documents

Provide the list of documents used to elaborate the technical content of the Integration Plan: System architecture description document, reports, minutes of meetings, others.

Reference	Title of the document

1.3.2 Applicable documents

Provide the list of documents entirely or partially applicable to the elaboration of the System Integration Plan form: standards, templates, descriptions of procedure, others.

Reference	Title of the document

1.4 Terminology: definitions and abbreviations

Provide the list of terms and their definition used in this document absent from usual dictionaries or used with a different significance from their usual significance.

Term	Definition

2 INTEGRATION OBJECTIVES AND CONSTRAINTS

Indicate the scope or limits of the integration; for example: integration of components, subsystems, system, incorporation of the system in its context of use, final validation on industrial site, final validation on operational site.

List or reference the system requirements that impact the integration execution.

List project constraints in terms of time (schedule), costs and risks.

List the contractual, legislative, standards, homologation and certification constraints to be respected.

3 INTEGRATION STRATEGY

3.1 Concerned components and interfaces

List the physical components to be integrated, possibly the hierarchical decomposition into (sub) systems and end components (not decomposable at the abstraction level concerned).

Indicate the interfaces between the components two to two.

Example of table:

Component 1	Physical interface	Component 2

3.2 Integration aggregates

Indicate the integration method(s) used.

List the aggregates, and for each aggregate, its composition in components, the assembly order (nr), the assembly technique (S: series or P: parallel), the reference of the aggregate definition sheet, see Annexe 8.10.

Example of table:

Aggregate	Sequence nr / Component	Assembly	Ref. Definition Sheet

For each aggregate, indicate the place of integration (where), the date (when), the assembly procedure (how), the required skills (who).

Example of table:

Aggregate	Place (site)	Date	Ref. Procedure	Skills

4 INTEGRATION ENGINEERING ACTIVITIES

List the integration activities and tasks. For this, instantiate for the system XX the generic activities, tasks and artefacts specified in the integration process in section 6.2.3 and section 6.2.5.

5 ORGANISATION AND RESPONSIBILITIES

Indicate the structure and composition of the integration teams, the respective roles.

Indicate the integration team's relationships with the other teams, as well as the assignment of tasks.

Indicate the responsibilities of each role.

6 MEANS AND TOOLS

Human Resources: for each role, list required skills, availabilities, related training.

Means and tools: for each means and tool, list the acquisition mode: purchase, leasing, development, subcontracting, its availability (date).

Integration site: for each site, list the place, infrastructures, platforms, etc.

7 INTEGRATION SCHEDULE

This is the schedule for the execution of the integration activities and tasks listed in paragraph 4 of this plan.

For the execution of the assembly of the components into aggregates, present a summary of the dates indicated in paragraph 3.2 of this plan.

8.10 Integration Aggregate Definition Sheet

Each sheet contains following information:

1 AGGREGATE

Identifier and name or title of the aggregate

2 ASSEMBLY

List of physical components that make up the aggregate

List of architectural characteristics and/or design properties covered by the aggregate

Verifications to be carried out before, during and after assembly of physical components

3 VERIFICATION AND VALIDATION ACTIONS

List of verification and validation actions to be performed on this aggregate

- *Verification of the functions of the aggregate*
- *Verification of input-output flows exchanged*
- *Verification of interactions (behaviour) between components*
- *Verification of temporal properties*
- *Expected results for each verification*
- *Drivers / launchers and/or stubs / caps required for each verification*

8.11 Integration Procedure

Each procedure contains following information:

1 PROCEDURE

Identifier and name or title of the procedure

2 PREPARATION OF INTEGRATION

Concerned aggregate(s): reference the Aggregate Definition Sheet(s)

List of means and tools necessary for assembly: harnesses, simulators, operator resources, enabling systems

Integration configuration: configuration of means and tools, aggregates, parameter values, necessary quantity entries, etc.

3 PERFORMANCE OF INTEGRATION

Orderly description of integration actions that includes:

- *Initial conditions*
- *Step by step assembly actions*
- *External stimuli or commands to be given by operators or tools to initiate or execute verification and validation actions (possibly refer to the descriptive sheets of these verification and validation actions)*
- *Intermediate check points and results to be recorded*

Description of results (after performance of the procedure) that includes:

- *List of results obtained from the assembly*
- *List of results obtained from performance of verification and validation actions*
- *List of encountered findings, observations, difficulties, anomalies during the procedure performance*
- *Synthetic conclusion: OK or not OK, decisions*